A Treasure Hunting Text

Ram Publications
Hal Dawson, Editor

New World Shipwrecks: 1492-1825
Comprehensive guidebook lists more than 4,000 shipwrecks; tells how to locate a sunken vessel and how to explore it.

Sunken Treasure: How to Find It
One of the world's foremost underwater salvors shares a lifetime's experience in locating and recovering treasure from deep beneath the sea.

Treasure Recovery from Sand and Sea
Step-by-step instructions for reaching the "blanket of wealth" beneath sands nearby and under the world's waters; rewritten for the 90's.

Ghost Town Treasures: Ruins, Relics and Riches
Clear explanations on searching ghost towns and deserted structures; which detectors to use and how to use them.

Real Gold in Those Golden Years
Prescription for happier, more satisfying life for older men and women through metal detecting hobby, ideally suited to their lifestyles.

Let's Talk Treasure Hunting
Ultimate "how-to" book of treasure hunting — with or without a metal detector; describes all kinds of treasures and tells how to find them.

The New Successful Coin Hunting
The world's most authoritative guide to finding valuable coins, totally rewritten to include instructions for 21st Century detectors.

Modern Metal Detectors
Comprehensive guide to metal detectors; designed to increase understanding and expertise about all aspects of these electronic marvels.

You Can Find Gold...With a Metal Detector
Explains in layman's terms how to use a modern detector to find gold nuggets and veins; includes instructions for panning and dredging.

Gold Panning is Easy
Excellent field guide shows the beginner exactly how to find and pan gold; follow these instructions and perform as well as any professional.

Treasure Hunting for Fun and Profit
A basic introduction to all facets of treasure hunting...the equipment, targets and terminology; totally revised for 21st Century detectors.

Buried Treasures You Can Find
Complete field guide for finding treasure; includes state-by-state listing of thousands of sites where treasure is believed to exist.

GOLD
of the
Americas

RAM
BOOKS

by
Jenifer
Marx

ISBN 0-915920-89-1
Library of Congress Catalog Card No. 96-68435
Gold of the Americas
© Copyright 1996
Jenifer Marx

First Printing, October 1996

For FREE listing of related treasure hunting books write:

Ram Publishing Company

P.O. Box 38649   •   Dallas, TX 75238

Gold of the Americas

Contents

Gold is tried with a touchstone,

Men, by gold.

Chilon (ca. 560 B.C.)

This book is dedicated to my Brothers,

Geoffrey Taylor Grand

and

Philip Frisbie Grant

and to the memory of our Grandfather

Henry Leslie Perez and his brothers,

who furnished the silver for my christening cup
from their mine near Tonopah, Nevada

Jenifer Marx

The Author

Jenifer Grant Marx is a writer interested in history and underwater archaeology. A graduate of Mount Holyoke College, she also studied history at the University of Florence in Italy. She was a member of the first U.S. Peace Corps contingent, serving in the Philippines from 1961 to 1963 where she worked on the islands of Negros and Mindanao. She subsequently lived in Europe, Africa, the West Indies and Asia.

Jenifer is married to Bob Marx, noted underwater explorer and author, in many of whose projects she plays an active part. They met in Jamaica where Jenifer assisted Bob who spent four years excavating the sunken city of Port Royal. They have three daughters and live in Indialantic on Florida's east-central coast when either or both of them are not traveling over the world.

Introduction

Tomorrow's
Gold

A small "calling card" made of gold is now hurtling through the vast and frigid silence of interstellar space. Attached to a derelict spacecraft, this golden plaque may someday introduce our civilization to worlds beyond the solar system. It will take 80,000 years for this golden greeting to travel to the nearest star...more than ten billion years to enter the planetary system of the nearest star. But, because its surface is gold, the six-by-nine-inch plaque can be expected to remain eternally shining and smooth, its message free of tarnish, scale or corrosion.

Etched on the plaque's surface are the figures of a man and a woman with their return address (Planet Earth). The inscription is engraved in scientific symbols that might conceivably be understood by extraterrestrial civilizations.

The plaque was affixed to the antenna support struts of Pioneer 10 when it lifted off from Cape Canaveral in 1972 on a voyage that has already taken it past the planet Jupiter toward a point on the celestial sphere near the boundary of the constellations Taurus and Orion.

Gold, the most precious of metals, was a fitting choice to carry the space age message. Since the dawn of history cultures around the globe have regarded gold as a divine substance connected with the sun. However, it wasn't gold's beauty or mystical associations that led scientists to use it for the first human message to leave our solar system. They chose gold because of its remarkable chemical and physical properties. Of course, it was these same characteristics that led the Ancients to regard the metal as a magic substance of divine origin.

The history of gold begins in remote antiquity. Without hard archaeological evidence to pinpoint the time and place of man's first felicitous encounter with the yellow metal we can only conjecture about those men who at various places and various times first came upon native gold. Bits of natural gold have been found in Spanish caves used by Paleolithic man about 40,000 B.C.

Yes, gold — a light that never dims — has held men in its thrall for thousands of years. No other metal has played such a consistent and dramatic role in the relatively short history of the frail creatures figured on the Pioneer 10 plaque. Gold is not essential for survival; yet, men have always craved it, subjugating whole nations and expunging civilizations and religions to get the yellow metal. No other substance has aroused such emotions or been invested with such power.

To see clearly the dichotomy of behavior gold has provoked we can trace the glittering golden thread through the annals of history and world literature...a source of joy and woe from the Golden Apples of the Hesperides to the Klondike poems of Robert Service. Lust for gold by men and nations has spurred exploration, furthered cultural exchange, developed international trade routes and hastened the development of technology.

Indeed, most nations of the Americas as we know them today can trace much of their history simply to the lust for gold. This most definitely includes the United States of America.

Part One

Nature's
Gold

N ature herself makes it clear that the production of
gold is laborious, the guarding of it difficult, the zest
for it very great, and its use balanced between
pleasure and pain.

That statement is as true today as when written by Diodorus
Siculus in the 1st century B.C. Some of men's most noble and lasting
achievements have been fueled by the precious metal or wrought
from it. Gold is a magic word. It conjures a kaleidoscope of images
from the myths and legends of every age and culture.

More than any other substance gold, imbued with an inherent
duality by man, has the power to inspire and to corrupt both savage
and sophisticate. "Gold is a child of Zeus," wrote Pindar in the 5th
century B.C., "neither moth nor rust devoureth it; but the mind of
man is devoured by this supreme possession."

Gold itself is simply a chemical element — deep yellow, soft
and dense — classified as a noble and heavy metal. The symbol
for gold, *Au* , is derived from the Latin word for "shining
dawn." This symbol, introduced in the 18th century, replaced
the prehistoric symbol which was representative of the sun.

Gold is too soft, heavy and pliable to be used in tools and
weapons. The Ancients learned quickly that gold knives take
no edge and gold plows make no furrows. Yet priest, ruler,
artisan, merchant, adventurer and citizen have long coveted,
sought and suffered for gold. Men have yet to find a universally

acceptable substitute for gold as a measure of value. Even in today's uncertain times the durability of gold represents a constant, defying both inflation and depression.

Humankind's highest attainments of craftsmanship have traditionally been in gold. Despite the ease with which the metal can be worked, it has never been responsive to an indifferent artisan. But in the hands of skilled men, the precious metal has been transformed throughout the ages into superb objects which mirror the extremes and diversities of history.

The lustrous metal, which is itself immune to foulness, has provoked wars, feuds, vendettas, false love and "dear divorce twixt natural son and sire." "What dost thou not compel the human heart to do, accursed greed for gold," wrote Latin poet Virgil at a time when already many thousands of slaves had died for it — men, women and children chained together and hacking, sifting, washing gold for Egypt's pharaohs since the fourth millennium. Throughout the stages of social evolution humans have always loved the gleam of gold.

Gold is the first element and metal mentioned in the Bible, where it appears in more than four hundred references. "A river flowed out of Eden," says the book of Genesis, "to water the garden and there it divided and became four rivers. The name of the first is Pishon; it is the one which flows around the whole land of Hav-i-lah, where there is gold; and the gold of that land is good."

Throughout history gold has brought suffering to those who labor to gather it, power and authority to those who hold it, and pride to those who wear it.

1 — *Symbol of Good and Evil...*

Magic Metal

Gold as symbol of both good and evil has been indispensable as simile and metaphor throughout the ages. Gold is the image of solar light, symbol of the life-giving principle. In Hindu doctrine it is "the mineral light," representative of the divine intelligence. Poets and writers have mined the golden vein for images of goodness, purity, truthfulness, inner worth, perfection, idealism and radiant beauty.

Yet oftentimes the metal has represented a potent symbol of evil. In ancient Greece Anacreon lamented, "In consequence of gold there are no brothers, no parents; but wars and murders arise from it. And what is worse, for it we lovers are bought and sold." In much of literature (and in history) gold is the seed of corruption and represents mammon, greed, lust and the erosion of humanity and principle.

"How quickly nature falls into revolt/When gold becomes her object," wrote Shakespeare in *Henry IV*, and in *Romeo and Juliet* , "There is thy gold, worse poison to men's souls/Doing more murder in this loathsome world,/Than these poor compounds that thou mayst not sell."

Wisdom and goodness, folly and corruption...opposite sides of the golden coin. Modern man knows that the malevolent powers ascribed to gold are imaginary but the emotional response to gold's aura of magic defies the rational. Rich in psychological overtones gold has been identified with the splendors and tribulations of human existence and so we identify golden ages, golden boughs, golden apples and golden dawns in poems, proverbs and romances as well as the golden rule, the golden mean and the golden number.

World-wide, gold has been used to strengthen the sanctity and worthiness of the spiritual ideal. The gods have demanded and priests have encouraged golden offerings as evidence of people's devotion.

Divine and secular authority have been bolstered by sumptuous displays of gold embellishing every conceivable surface including ornaments, garments, fabrics, embroideries, cosmetics; façades, walls, floors, ceilings, domes and other architectural elements; furniture, wood and metal work, ceramics, glass, leather, stone, paper and shell.

A famous dictum advises women that they can never be too rich or too thin. There also seems to be no such thing as having too much gold. The thirst for gold has seldom been slaked. With a literal and horrible justice, the King of gold-rich Parthia had molten gold poured down the throat of the defeated Roman general, Marcus Crassus, who had invaded Parthia for its gold in the 1st century B.C. In medieval Russia counterfeiters were similarly punished with molten gold funneled down their gullets. Sales of gold jewelry on today's television shopping networks prove that late 20th-century Americans crave it every bit as much as the Ancients. The big difference is that gold was once the property of rulers and gods but today almost everyone has a gold ring or chain.

First Encounter

Imagine, if you will, Stone Age men wandering in search of roots, berries and shellfish in a world before memory when gold lay exposed in rain-swollen streams or on ground laid bare by the elements. A man following a watercourse in search of food would have been struck by the gleam of a small tumbled nugget deposited by gravity in the shallows. Picking it up, he would marvel at its glow, the smooth warm feel of it in his palm, and its extraordinary heaviness. He took it home to his shelter. The others in his small group admired it. Perhaps he was content to keep the dazzling stone as it was. But he may have seen the buttery mark it left when rubbed across a rough surface or noted how easily a piece of bone or even his fingernail could score it. At some point he began to regard it as a kind of talisman, unaware as yet how easily it could be shaped by blows from his stone hammer. Eventually he learned to cold-hammer native nuggets into simple flattened roundels. By that time he may have made an inarticulate association between the sun and the bright lump of stuff that warmed in his hand.

Experts are divided. Was gold the first metal man worked, or was it copper? Gold, copper and meteoric iron (the only iron known to the Ancients) are the only metals found in relatively pure form in the native state and the only metals with significant color in their free state.

Thousands of years before it became the first metal to be smelted from its ore (about 4300 B.C. in western Asia and about 3800 B.C. in Egypt, Mesopotamia, Syria, Persia and India), copper was found as varicolored chunks, thin laminate sheets and branch-like pieces of pure metal. These broke off from veins of copper that ran through a matrix of basic stone and ores.

The oldest metal artifact found is a copper pendant made some 9,500 years ago in today's Iraq. Copper is soft enough to shape by cold hammering yet hard enough to take a fair cutting edge. Iron, next to aluminum, is earth's most ample metal, but cannot be worked without extraction from its ore. This didn't happen until thousands of years after men had learned, by trial and error, to smelt copper, which melts at a much lower temperature than the 3,650 degrees Fahrenheit needed to smelt iron. Once man accumulated the relatively sophisticated technology to smelt iron, it gradually replaced copper and bronze as the basic metal for implements and weapons.

But primitive man knew and cherished nearly pure iron found on the earth's surface in the form of siderites, metallic meteorites, which ranged in size from pebbles to gigantic masses of several tons. Long before the Sumerians dubbed it "heaven's metal primitive man recognized its celestial origin. Early smiths managed to chip bits from larger pieces and work them into amulet and ceremonial objects using stone tools.

By 1200 B.C. iron was available to the farmer and soldier in the West, but no one can explain how smiths made the few iron artifacts found at the Royal Cemetery at Ur dating from prior to 2500 B.C. or at the Alaça Huyuk site in Anatolia where archaeologists unearthed a third millennium B.C. iron pin with a gold head and a fragment of a crescent-shaped plaque. Two sites in Egypt dating back 6,000 years yielded iron beads, an iron knife and an amulet of silver and meteoric iron.

Gold has never been a utilitarian metal, although men have made it into bowls, knives, fishhooks, and scrapers. Some archaeologists contend that copper was worked before gold. A few believe gold was not at valued by prehistoric man. This seems unlikely. Neither copper nor meteoric iron have a hue or sheen like that of gold nor are they as malleable. The first people who used gold were dust long before the advent of written or oral history, but because gold is imperishable, archaeologists may someday find and date gold artifacts that predate the oldest copper objects. There is no argument that by the time nomadic hunter-gatherers began the transition to agriculture which required a settled pattern of existence gold had begun to be highly valued.

The Neolithic age was characterized by protofarmers of northern Iraq and Persia, who were leading a farming and stock-raising existence by about 6000 B.C. Over the next thousand years protofarming spread to the Nile Valley, the coastal areas of Palestine, the upper basin of the Euphrates, northern Syria and central Persia.

People living in small permanent settlements made pottery, wove cloth, reared flocks and cultivated grains such as barley and wheat. Those first farmers became aware of the crucial role the sun played in ripening the grain. They worshiped the sun because it was the source of life-giving light. What more appropriate offering to the sun than gold...the gleaming "sun stuff?"

The First Metalsmiths

With a settled pattern of existence came specialization. The descendants of the men who had been most skillful at making flint and stone choppers and scrapers became the first metalsmiths. They practiced a craft that required them to devote full time to it. They were unusual men who experimented and tested. Little by little they puzzled out the techniques that yielded the first copper and gold artifacts.

They made tools, weapons and religious pieces. Metallurgical historian Cyril Stanley Smith writes, "Nearly all the industrially useful properties of matter, and ways of shaping material, had their origins in the decorative arts...the first suggestion of anything new seems to be an esthetic experience."

More than a million dollars worth of gold coins recovered in 1964 from one of the 1715 Spanish shipwrecks near Fort Pierce, FL.

The prehistoric toolmakers who evolved into the first specialized metalsmiths were regarded with awe. They practiced an art surrounded with an air of mystery and secrecy and could perform feats which amazed their primitive fellows. It was fortunate for the earliest smiths that they had such a superb medium to work in, a metal of such wonderful color and soft rich texture that men through the ages have fashioned their greatest artistic offerings in it, a metal which lends itself to shaping and surface decoration and then remains forever whole and beautiful.

The most obvious appeal of gold is its loveliness; the most significant is what alchemists in the Middle Ages termed its nobility. It is a chemically inert element, and neither air, moisture, nor common acids or alkalis can mar its beauty, which is why gold ornaments fashioned six thousand years ago are lovely today long after more recent artifacts have turned to dust. Gold's link with eternity and its perpetual luster led the Ancients to think of it literally as sun stuff, and imbue it with an aura of sanctity and magic.

Not only Neolithic man but Sumerians, Egyptians and New World peoples like the Incas and even contemporary primitive groups have regarded gold as the fruit, seed, sweat, tears or excrement of the sun. Traders took advantage of this hallowed regard for gold to establish an economic role for the metal.

The hoary phrase *"withstanding the acid test"* refers to gold's stability in most acids and originated before it was widely known that gold, despite its chemical snobbishness, does react with chlorine and other halogens and can be dissolved in a heated mixture of nitric and hydrochloric acid. This mixture was called *aqua regia*, royal water, by the alchemists because it attacked the royal metal as scale, rust and time could not.

Primitive metalsmiths were delighted with golden lumps which could be hammered without breaking and shaped easily without

It was the dream of ancient alchemists and sorcerers to develop "magic" formulas that would enable them to transform base metals into gold.

heating. In relatively pure form gold has unrivaled qualities of malleability and ductility. One troy ounce of gold, for instance, can be drawn into a hair-like thread 50 miles long. The same amount of gold, no larger than a sugar lump, coats more than a thousand miles of silver or copper wire. Before the second millennium B.C. wire was made by rolling out short lengths of gold welded together. In Egypt in the second millennium gold, copper, and bronze wire were drawn through a perforated stone.

Gold can be beaten or rolled into translucent sheets so thin they transmit a greenish light and so delicate they can be moved and straightened with a light breath. One troy ounce can be beaten to a golden film covering more than 100 square feet. It would be so thin that 1,000 sheets would be needed to make up the thickness of this page. A three-inch cube can be made into a sheet covering an acre.

"Goldbeating" (the making of gold leaf) was well known in the pre-Biblical world. It is one of the very earliest of crafts and one of the very few to have resisted mechanization. Early gold beaters prepared gold leaf in much the same way it is made today, hammering a small amount of gold between sheets of parchment or specially prepared animal membrane and burnishing the gold with gems, semiprecious stones, boar's or dog's teeth. From remote times gold foil was used in Asia and the West to ornament wood, pottery, and textiles.

The Roman historian Pliny wrote that a small quantity of gold could be reduced to 750 leaves each four digits (fingers) square. Ancient Rome glittered with gold which was applied to the façades and interiors of palaces, temples, and public buildings. The Egyptians made the most spectacular use of it in the overlays of gold leaf on furniture and royal mummy cases.

Five thousand years ago the obelisks raised by sun worshipers as "conductors" to transmit the life-giving, harvest-ripening rays of the divine fire into the earth and fructify it were often sheathed in gold. The Egyptian goldsmiths were not above practicing golden deception. They were the first to coat objects of inferior metal with gold leaf and pass them off as the real thing, a practice which continues even today.

The Greeks, who taught the art of gold leaf to the Romans, not only gilded masonry, sculpture and wood but also fire-gilded metal by applying a gold-mercury amalgam to it and then driving off the mercury with heat. Today gold leaf offered by the devout gleams from thousands of Asian temples such as Rangoon's Shwe Dagon. In the West the dim interiors of countless churches are illuminated by the glow of gold from altars, architecture and vestments.

Gold like that which shone from Mesopotamian ziggurats now shimmers from gold-plated roofs, the windows and exterior walls of fast-food shops and office buildings. Gold is an ideal coating material that can be manufactured at relatively low cost to protect against the ravages of weather and time. Gold shining from the spires and domes of public buildings also tends to proclaim their importance.

Soluble in oil, gold is used as a film as thin as five-millionths of an inch to ornament glassware, ceramics, paper and plastics. It beckons the consumer from decorations on perfume bottles and shaving-lotion containers. Gold ceramic tiles reflect sunlight on the roofs of motels and restaurants. A sandwich of laminated glass containing a film of gold less than two-millionths of an inch thick is used for the windows of modern aircraft. The gold prevents icing or misting, reflects glare and guards against harmful rays of the sun.

Powdered gold in a ground of chalk, marble dust, and an animal glue or *size* has been used for millennia to gild large and small surfaces, to embellish leather, manuscripts and miniatures and for countless medicinal and cosmetic purposes.

At first gold was worked as it was found. Much later men learned to combine it with other metals to form alloys which increased the element's strength and changed its hue. Gold can be combined with varying amounts of silver, copper, nickel, zinc, palladium, platinum and iron to produce a wide range of color including yellow, red, orange, green, blue, white and purple. In Tang dynasty China an alloy was made with minute traces of iron which turned the gold violet when heated. A rare form of gold appears black because of the addition of bismuth. In nature gold is often alloyed with silver. This natural alloy, called electrum, was use by the tyrant Gyges of Lydia in Asia Minor in the 7th century B.C. to make the world's first coins.

Physical Properties

Gold is extremely heavy: one and a half times as heavy as lead and twice as heavy as silver. The lovely sheen and rich texture of a small gold ingot belies its density and weight. The metal is so compact that a cubic foot of pure gold would weigh about half a ton. The heaviness of a gold coin or ornament heightens its sense of value.

Gold melts easily, at 1,064 degrees centigrade, which was well within the temperature range of ancient cooking fires. When it boils, the metal remains yellow but gives off an unusual purple vapor which must have added to the goldsmith's reputation as miracle worker.

No one is sure how long ago gold was worked. However, by the time the Royal Cemetery at Ur was created ca. 2,500 B.C. Sumerian goldsmiths had already mastered most of the technical processes required by goldworking, including casting in open molds and lost-wax casting. They had long since discovered how to anneal gold by heating, quenching and beating over and over.

The first appeal of gold was esthetic and its first applications were of a magical nature. However, human vanity being what it is, men soon desired gold adornments for themselves, and goldsmiths fashioned ornaments attesting to the rank of the wearer in life and in death. Much of what we know of preliterate life comes from ancient mankind's treatment of the dead. Around the world, men, women and children were laid to rest with comforting objects as simple as flower bouquets and as elaborate as the awesome armies of warriors found recently in tombs in China. Many thousands of years later as these artifacts are unearthed they illuminate the past in a unique and moving way.

An Economic Medium

In the evolution of barter economy into more sophisticated economic systems gold proved to be the one substance universally accepted as payment for goods and services. It could be divided without damage, stored without deterioration and conveniently transported in small amounts to concentrate a great deal of value. It could be easily concealed or ostentatiously displayed as status-enhancing jewelry.

Thus from earliest times gold has moved through history in many guises providing a magnificent medium for the artisan, artist and jeweler...a fitting offering to gods, rulers and lovely women...a universally prized store of wealth and motive for both heroic and dastardly deeds.

But it is not gold's beauty, mystery, or rarity that recommends it to modern technology. Gold is valued in industry, space, and defense programs because of its resistance to corrosion and change, its high conductivity of electricity, and its superior reflective

14

properties. Gold conducts electricity better than any metal save copper. A minuscule amount of liquid gold on a printed circuit can replace miles of wiring in computers. It is found in the minute circuitry of transistors, undersea cables, telephone equipment, radar, televisions and calculators. A plating as thin as three- or four-millionths of an inch is enough to protect a component from deterioration.

Space age applications extend far beyond the unique message plaque which may (or may not) be deciphered in the distant future. Gold reflects up to 98 percent of incident infrared radiation. It is such a fine reflector of heat and light that an almost nonexistent film on the heat shield of a rocket engine can protect the fragile instruments within from the searing heat generated by the lift-off thrust and the reentry into earth's atmosphere. Edward White, the first American to walk in space, had a gold coating on the umbilical cord which connected him to the Gemini spacecraft. The visors on astronauts' helmets are gold-coated to filter out both infrared and ultraviolet radiation. Gold sheathing on the tiny television camera and other instruments on the moon buggies blunted the sun's noxious rays. Gold-coated nylon was used as a protective shroud for the radio, propulsion, and guidance systems of the Gemini craft.

Medicinal Applications

Gold has found widespread application in dentistry and medicine. Because it is malleable, acid-resistant, nonpoisonous and chemically tasteless, dentists today find it ideal for fillings, inlays and caps...as did the dentists of antiquity. A 4,500-year-old gold bridge was found by archaeologists in an Egyptian tomb near El- Quatta. In areas like the Philippines where a flashy display of gold is considered in good taste, it is not unusual to see teeth sheathed in gold or sporting decorative gold inlays. The smile of a winsome lass may reveal a small golden heart, flower, or bird inset in the lateral surface of a front tooth.

The medical uses of gold are an ancient heritage. Gold was considered a potent curative force in early medicine. In China, gold leaf was considered the most perfect form of matter and a salve of gold was the most powerful of Chinese pharmaceuticals. The unalterability of gold linked it with eternity leading to the belief, incorporated in alchemical practice, that gold could give renewed or prolonged life to the body.

15

In the West *aurum potabile*, elixir of gold, was widely prescribed for a multitude of ills. Ointments, gold salts and gold pills have been taken through the ages for the treatment of many disorders including tuberculosis and syphilis. Writing during the Roman era, Pliny the Elder mentions salves of gold prescribed for ulcers, fistulas and hemorrhoids. Gold tablets have been touted for their energizing powers even to the present day, and an American folk belief persists that a sty disappears when rubbed nine times with a wedding band.

Astrological healers claimed that gold medicines ingested under certain signs of the zodiac would cure appendicitis. Several decades ago millions of gullible alcoholics submitted to the costly and painful Keeley cure; a nonsensical treatment that involved the injection of gold chloride as a first step. In recent years injections of radioactive colloidal gold have aimed at alleviating the pain of arthritis. Radioactive isotopes of gold have been used to irradiate cancerous tumors. Gold-silver alloy plates help to repair skull injuries, and gold has been employed in the treatment of certain kinds of ulcers, bums and some nerve-end operations.

Tongue in cheek, the Italian Vannoccio Biringuccio attested to the therapeutic effects of gold in his 16th-century treatise on metallurgy. "Indeed as a medicine it is beneficial to certain illnesses. Nature with her own virtue has endowed gold, as a singular privilege, with power to comfort weakness of the heart and to introduce there joy and happiness, disposing the heart to magnanimity and generosity of works. Many learned men say that this power has been conceded to it by the benign influence of the sun and that for this reason it gives so much pleasure and benefit with its great powers especially to those who have great sacks and chests full of it."

Indeed, nature's gold has never failed to gladden hearts and fire the imagination. Those who have not had sacks of it have aspired to, and those who have had chests of gold have yearned for still more. There has never been enough gold to satisfy the appetite which seems to grow with feeding. Gold still enhances those who wear it, comforts the anxious and throughout history gold has kept its value. An ounce of gold, according to the Bible, bought 350 loaves of bread in Babylon during King Nebuchadnezzar s reign. Today…in spite of inflation, it still buys 350 loaves.

2 — *Gold is Forever...*

Eternal Symbol

No bit of matter is intrinsically precious. Yet through the ages man has endowed countless substances — animal, vegetable and mineral — with value. Shells, totem poles, animal skins and beads have all served as a measure of wealth at various times and in various places. There are cultures where boar tusks or whale baleen purchase a bride that gold could not. But gold has a universal appeal and a host of associations that shells and teeth can never match.

As a symbol of power both divine and secular and as a measure of value, gold has played a leading role in shaping the human experience. The soft yellow metal has been a major catalyst in the ebb and flow of civilizations. Provoking the rise and fall of empires, states and individual men and women, deathless gold has emerged undiminished from the dramas it precipitated.

Because gold endures, it accumulates. A large part of the gold that passed through the hands of the Mesopotamians, Egyptians, Persians, Greeks, Scythians and Incas, to name but a few of the peoples who had an intimate relationship with the precious metal, is still around. Some estimates run as high as 80 per cent.

But...gold, the indestructible, has its Achilles' heel. It is very easily melted. During periods of turmoil, to cover theft or simply to keep up with the vagaries of fashion gold jewelry, vessels, coins, ingots and idols have all been tossed into the melting pot to emerged in new guise. Not even royal crowns escaped. In fact, monarchs often had a hard time keeping their regalia out of the pot and out of hock. In England magnificent gold objects that had survived Henry VIII's seizure of ecclesiastical treasure were wantonly melted by Cromwell. In India today women periodically have their gold bangles altered to conform to current styles.

Golden grave goods, buried with the wealthy, were made to placate the powerful gods and comfort the deceased on their journeys into the infinite. But even in ancient Egypt gold consigned to the tomb was often dug up almost immediately by organized teams of robbers. Gold offerings and idols were looted from sanctuaries or seized as war booty. Melted and recast in the taste and temper of the times, infinitely beautiful gold passes through successive ages in a multitude of forms. Gold that bedecked a Mesopotamian princess or graced an Inca palace may now be part of the Pioneer plaque soaring into the universe where it originated eons ago when the earth was formed.

The allure of gold has something of the mystical about it; the links with economics and political power were forged long after it achieved preeminence as a divine metal. The religious association between the sun and gold contributed to the adoption of gold as the royal metal and an incorruptible medium of exchange.

Possession of gold enhanced authority, and in the days when rulers were considered actual descendants of deities or later when they were recognized as fully human but ruled by divine right , gold reinforced a ruler's aura of authority and sanctity. Even today men and women frequently gain power through "sanctissima divitiarum majestas," the most sacred majesty of wealth.

Many cultures regarded gold as a mystical element of celestial origin. The Scythians in the 7th century B.C. had a saga of the sacred gold which fell from heaven. A great solar myth underlies virtually all ancient mythologies. Gold, the "seed of heaven," the "excrement of the gods," the Hindu's "mineral light," and the sun with its nourishing heat and light were aspects of the same basic principle.

Sir James George Frazer's Golden Bough mentions that until recently central European gypsies believed in a manifestation of the life-giving principle of light-and-gold. They called it the fern seed, and it figures in the myths and ceremonies of many European areas, although today its form has been altered almost beyond recognition. For thousands of years parasitic mistletoe which grows in trees, particularly oaks, has been ritually linked with sun and gold, fire and light. People believed that if fern seed and mistletoe were gathered at the solstices the vegetable-gold emanations of the sun's fire would reveal earthly treasures of mineral gold.

Men believed gold grew as plants and could beget offspring. In A.D. 1540 an essay on "the Generation of Metals" averred that "in some places of Hungarie at certain times of the year, pure gold springeth out of the earth in the likeness of small herbs, wreathed, and twined like small stalks of hops, about the bigness of a pack of thread and four fingers in length." Centuries later some Europeans still believed that "the tailings of abandoned mines became so enriched as to be workable at a profit, the amount of gold thus obtained being proportional to the interval of time elapsed."

In fact, although methods of gold refining remained unchanged for millennia, men were sometimes able to reprocess old tailings and extract small amounts of gold. In the late 19th century several English mining concerns processed ancient tailings piled on the Egyptian desert where they had been left by slave miners five thousand years before. They discovered that the extraction process of the Ancients had been such that almost no gold remained. They had somewhat better luck in the mines themselves, finding veins and pockets of ore in the twisting, ruined galleries some of which reached six miles inland from the Red Sea shore. The ancient diggings yielded vast numbers of human bones and $3 million in gold before they were abandoned in 1921.

There are areas, particularly in Southeast Asia, where gold mining is surrounded with ritual and the gold itself considered a living thing. The Dyaks of Borneo believe the precious metal has a soul which will avenge itself on those who wrest it from the earth. Miners in parts of Sumatra formerly prospected for gold because they believed they were stealing the precious stuff from the spirits who would punish them if they talked. If the miners brought tin or ivory to the mining site, they believed the spirits would make the gold vanish. In neighboring Malaysia tin miners traditionally regarded tin as alive and growing, sometimes in the form of a water buffalo which wanders from place to place.

Goldsmiths were always accorded high status. Among certain West African tribes only a chief may be a smith. And in parts of Sumatra and Borneo goldsmiths were considered neutrals in war and allowed to pass unharmed through hostile territory. The goldsmith of the Karo Batak in Sumatra practiced a sacred craft; before starting work he prayerfully offered his blood, heart, liver and lungs to the spirits. He considered his tools as animate objects capable of changing their names and functions. Thus, he used a "secret" language when working. It was customary for a son to follow his father because if he didn't the tools would "ridicule" him. Goldsmiths' tools could not be sold without causing great harm to seller and buyer; they must be inherited.

There is no rational explanation for gold's allure, but its history, the best documented of any metal's, reflects the emotional response it has aroused. The Greek playwright Euripedes noted, "There is a saying that gifts gain over even the gods. Over man, gold has greater power than ten thousand arguments." The golden thread is woven into the fabric of human history...in magic, religion, philosophy, poetry, art, politics, economics, science, medicine, industry, fashion, romance and conflict. There is scarcely a time or place in which the precious metal has not figured in some way. It has baited the hook of charlatans, con men and counterfeiters for 60 centuries.

Supplies of gold have never been adequate to meet demand, although there have been moments of unprecedented and even unsettling abundance such as when Alexander conquered the golden East and Persian gold flowed into the Hellenistic world or in the years following the 1849 discovery of gold at Sutter's Mill in California. There have been eras of great scarcity too, particularly following the Fall of Rome when the barbarians didn't know how to work the rich Roman mines.

For thousands of years men strove to manufacture the precious metal. Alchemists have too often been lumped together with the rogues who, in one way or another, have taken advantage of man's greed for gold. Alchemy, practiced in China and the Middle East centuries before Christ, was ostensibly the search for the philosophers' stone, which would transmute base metals into gold and confer eternal life. The alchemists, traditionally regarded with ridicule, did far more than mix dragon's blood and gazelle urine in the smoking retorts of their dank laboratories.

Undoubtedly many of them were hopeless dreamers or even swindlers, but the empirical experiments of such men as Geber in 8th-century Baghdad, Robert Boyle in 17th-century England, and countless other dedicated alchemical scientists laid the foundations for much of modern chemistry, optics, physics, biology and medicine. Among the ranks of Hermetic scientists were popes, kings, doctors and even women as well as devotees of arcane and magicians who made a career out of relieving wealthy clients of their gold with the promise of doubling it in their laboratories.

In the 15th and early 16th centuries, the paucity of gold, which kept the alchemists struggling in their reeking dens, propelled men across uncharted seas in search of new sources. King Ferdinand of Spain sent his conquistadors westward to "get gold, humanely if possible, but at all hazards get gold." But the cost of getting gold has always been fatally high to some. Before the onslaught of a small number of Spaniards in chain mail who had fire-belching sticks and rode fleet horses, animals never before seen in the New World, the Inca, the Maya, the Aztec and the Carib succumbed. High cultures and low were extinguished. Indian slaves died like flies.

Ironically the conquistadors were no doubt unaware that in their homeland Roman slaves had routinely been worked to death deep in Spanish gold mines. Even earlier in Egyptian mines of the Kush, slave laborers were considered expendable because it was more costly to bring in supplies and water for the miners than to bring in replacements. In the 1890s the hapless Ona Indians of Tierra del Fuego were shot at like rabbits by the rough men who flocked to that desolate, wind-swept tip of South America in a short-lived gold rush. And very recently in the wilds of Brazil a gold boom has taken a toll on warring miners and an even greater toll on hapless Amazonian natives.

The world shrank before the exploring traders and intrepid gold seekers who pushed beyond the known world's frontiers into harsh lands and across dread seas filled with nightmare and superstition. Greek chroniclers writing in the first millennium B.C. described hostile climate and terrain inhabited by gold-guarding birds of prey and gold-digging ants as large as foxes, but nothing stopped the gold seekers. There has not been an age without its Eldorado, a fabulous golden land or city shimmering just beyond the ken. The Chinese, Japanese, Indians, the inhabitants of the Near East and the Mediterranean...all had tales of legendary kingdoms with inexhaustible golden riches and often thought to be the abode of immortals where one could find everlasting peace and joy.

For thousands of years one goal of the quest was Ophir, site of King Solomon's mines. The location of Ophir had been a matter of much speculation in antiquity. The golden land's position shifted from India to Arabia to Africa, and as men explored those zones without finding it the search focused on the Caribbean, Greenland and South America. Columbus set out to find the golden land of Cipango (Japan) Marco Polo had written of but ended in the Caribbean. In 1976 United States and Saudi geologists announced they had located Solomon's mine in the Saudi Arabian desert.

Medieval Europe was riveted by the tale of Prester John, a Christian potentate who presided over a mysterious Asian kingdom drenched in gold. The Pope and European kings actually received letters alleged to be from the powerful ruler. His legend persisted through the centuries, and many expeditions were launched to find the man who would be a valuable Asian ally in Christendom's mortal struggle with the infidels.

Asia was combed for centuries but no such golden kingdom was discovered. Men whose dreams of gold could not be quenched then turned to Africa to find Prester John and particularly to Ethiopia, where there was indeed a Christian ruler and a great deal of gold. Not everyone was satisfied the great John had been found there. Sir Walter Raleigh, not immune to the lure of legend, sought Prester John in the jungles of Guyana, pursuing his golden dream to ultimate ruin.

Spaniards, Portuguese, and even Germans before him had plunged into trackless jungles, scaled mountains and braved unknown waters to find Eldorado. The quest took them across North America and throughout Central and South America. Coronado trekked across the great Southwest in search of Quivira; Balboa looked for Davaive, a Panamanian cacique (chief) who reportedly had scores of workers whose only task was to melt gold for his dazzling treasure hoard.

Quivira, Eldorado, the Seven Cities of Cibola. They had different names, hazy locations, but all were equally elusive until Pizarro in 1532 found the golden Inca kingdom in Peru which surpassed even the most glittering dreams.

In our 20th century, when every bit of land and water has been probed and labeled, the quest still goes on. In the early part of this century the Krupps, German industrialists, mounted a large and costly expedition to penetrate the interior of Brazil in search of a legendary lost city of gold. They organized with characteristic efficiency but were encumbered by too many people and supplies. In the interior they ran out of food, and the animals could find no forage. The Krupps admitted defeat, but there are many who still believe that gold lies waiting in the heart of the Amazon jungles.

Gold seekers throughout the ages have always been unusual men, optimistic, adventurous, and resourceful. Their quest has never flagged, though it shifted from one area to another. One of the first documented prospectors was the Phoenician Cadmus who is credited with introducing the beginnings of our written language. In the 14th century B.C. he went from Canaan north to mainland Greece, where he found gold and according to Strabo "carried there the alphabet and other germs of civilization."

The Phoenicians were unsurpassed masters of maritime trade in the ancient world. They sailed through the Pillars of Hercules, out of the familiar Mediterranean. They ventured down the coast of Africa, where they engaged in the famous "silent trade" for gold, and up to the British Isles for tin in the second millennium B.C.

Pursuing the will-o'-the-wisp of gold, men have pioneered new lands, building cities in the wilderness, spreading and blending elements of culture, religion and technology. Success has only served to inflame the desire for more of the stuff. To get gold men have inflicted incredible suffering on others and subjected themselves to unbelievable hardships. To reach the Klondike, for example, men voluntarily undertook the most appalling peacetime advance in history. The lure of gold has drawn men across continents, oceans, deserts and mountains to lonely deaths and only rarely to a bonanza.

The sea today offers the greatest potential source of gold. Billions of dollars' worth of sunken gold, the accumulation of centuries, lies on the bottom, much of it beyond the reach until late 20th century developments in deep-sea technology . The greatest amount of ocean gold, however, is in colloidal form and awaits harvesting when and if an economical way is found to extract gold from sea water.

Today when there are so many substances, many of them unknown to antiquity, to catch one's fancy, gold is still esthetically pleasing and highly valued. Until fairly recently it was undisputed as the preferred way to display wealth, reflect divine grace and confirm worldly power. Because of the nature of the metal and the practice, until Christian times in the West, of burying the dead with gold artifacts, we can look today on the work of thousands of nameless who crafted objects of enduring beauty and in some eras of crashing vulgarity. Sometimes a legacy of gold ornaments, distinctive and lovely, their luster undimmed by the passage of time, is all that remains of a culture. In every age the eloquence of goldsmiths' creations reveals mankind's high regard for the rich yellow metal.

Grave gold delights the senses as rusted iron can never do. Gazing, for example, at gleaming Scythian pieces, rich in texture and bold in conception, one feels an empathy for the vigorous nomads who roamed the Eurasian steppes with their horses and cattle two thousand years ago. Their magnificent grave goods are all we know of the Scythians' material culture, but they bring to life the descriptive pages of the 5th century B.C. Greek historian Herodotus.

The ancient practice of furnishing the dead with precious objects as well as humble but favored possessions has proven a boon to the modern world. In spite of the conquistador's destruction of New World gold and silver-works, tombs violated in the 20th century have yielded treasures which make viewers marvel at the consummate artistry and skill of American craftsmen. Most of what we know of the remarkable goldwork of ancient civilizations, such as the Egyptian, Sumerian, Mycenaean, Etruscan, Iberian, Scythian, Celt, Roman and Greek, comes almost entirely from archaeologists' legitimate excavations of tombs. Unfortunately for the archaeologist and historian who want to put together a picture of ancient civilizations, grave looting has been going on as long as men have been buried with their gold. Egyptian papyri of about 1100 B.C. tell of official investigations into thefts from royal tombs and other sacred places along the west bank of the Nile at Thebes. The robberies involved political scandal, the involvement of artisans and minor bureaucrats.

Eventually it was deemed necessary to transport royal mummies to secret tombs in the Valley of the Kings at Luxor. However, small gangs continued their depredations and managed to despoil virtually every tomb. One of the few which remained unscathed was that of the minor prince young Tutankhamen, whose splendid tomb was found by Howard Carter in 1922.

In the 18th century, when things Etruscan were in vogue, Italian noblemen on whose vast estates Etruscan tombs were located hired peasants to plunder the tombs. Such robbers often routinely crushed underfoot anything that was not precious, and when a market developed in ceramics the peasants were told to break all pots that were not unique. This inflated the market value of a few ceramics, making fortunes for the landowners, but outraging historians.

There are still treasure caches in many parts of the world, and still thousands of grave robbers whose only concern is the clandestine removal of salable items. Despite government efforts to stop them, these men, encouraged by unscrupulous collectors and museums which are not choosy about the provenance of artifacts, probe the ground from Turkey to Central America. In Panama the guaceros' illegal trade is so well organized they have even formed a union. Until a few years ago grave gold in Panama was sold under the counter, melted and used by local dentists. Today the there is such a good market for antiquities that forgers around the world turn out endless imitations which range from crude to almost perfect.

The greatest artistic loss of gold workmanship was probably the melting of the incredible golden masterpieces seized as Spaniards conquered the New World. They left us detailed accounts of gold-plated buildings, trees, flowers, animals, and birds all life-size, fashioned of gold. With no hesitation the conquistadors consigned the genius of a dozen cultures to oblivion. The sacred gold was thrown into the crucible and emerged as the bullion and specie which nudged Europe into an unprecedented inflationary spiral. In the long run the gold gods exacted a full measure of revenge on Spain, leaving her weakened and impoverished.As a monetary metal, gold, after thousands of years of service is being phased out by economists and politicians who, like Keynes, view it as a "barbarous relic" and seek to strip it of hallowed mystique. Returning to the gold standard is no longer possible nor desirable, but monetary gold has its passionate advocates, among them the French, the Russians and certain conservative American candidates who argue for a role for the metal. The debate is not yet over. No currency has ever equaled the record of gold's stability, although the British sterling throughout most of the 19th century and the United States dollar after World War II were as good as gold.

Germans who pushed a baby carriage full of depreciated marks to the store at the end of World War II to buy one loaf of bread or South Vietnamese holding their savings in a currency wiped out after the United States withdrawal in 1975 would have been far better off holding gold. Traditionally, in most of the world, gold has been excellent insurance for volatile times.

25

In times of uncertainty gold keeps its value and represents a form of anonymous wealth which can be transferred without red tape. Gold quietly crosses borders; no questions are asked of it beyond assaying for purity, which is easily done. Gold is still occasionally used as payment for espionage and intelligence work.

Gold resists being ranked with soybeans and pork bellies on the commodities market. Although it has been dethroned from international monetary supremacy, it still remains the ultimate acceptable form of payment between individuals and nations, whatever their differing philosophies. While experts continue to debate what part gold should play in global economics, anxious people in Argentina, Indonesia, India and Brazil continue to trust gold for the preservation of capital. Most countries continue to treat their gold reserves as an asset of last resort despite the fact that as Robert Triffin of Yale pointed out, "Nobody could ever have conceived of a more absurd waste of human resources than to dig gold in the distant corners of the earth for the sole purpose of transporting it and reburying it immediately afterwards in other deep holes, especially excavated to receive it and heavily guarded to protect it."

Some seven thousand years of experience say gold is forever.

Over

Artisans of Peru's Moche culture, 300 B.C.-200 A.D., crafted the intricate golden crown, bottom, and the beautiful necklace, top.

Facing

This exquisite pomegranate from the ancient Greek city of Mycenae, circa 1400 B.C., was found on the Mediterranean island of Crete.

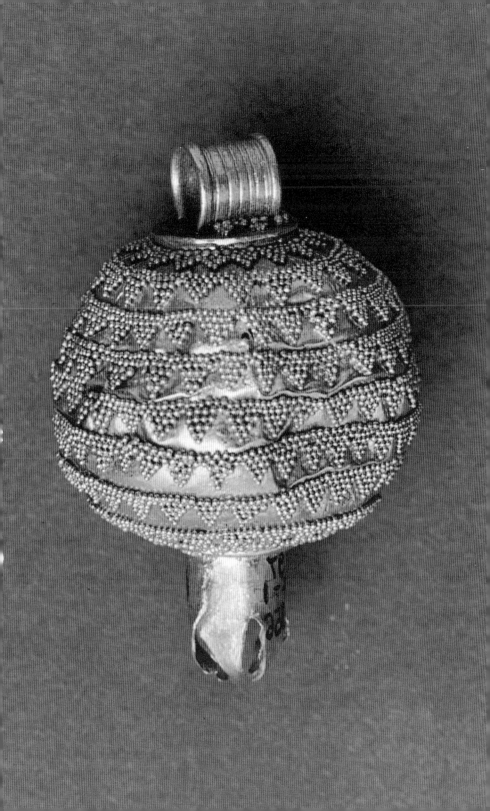

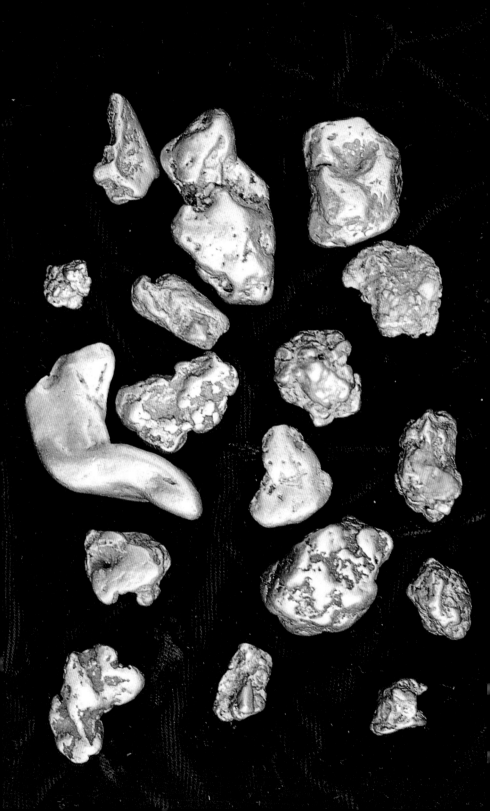

3 — *From Fiery Star Dust...*

Gold's Origin

G old is almost everywhere...in deer antlers, polar ice, human hair, plants...even the clay underneath Philadelphia and the buildings of Johannesburg. Yet Nature seldom sowed the most coveted of metals as prodigally as she did in King Midas' Pactolus River or along California's Mother Lode. For the most part gold is distributed around the globe in invisible and often inaccessible form. The estimated gold production of the past 6,000 years, although a colossal amount, pales beside the 10 billion tons of gold still suspended in the oceans and the 50 billion trillion troy ounces present in the sun as hot gas.

The oldest of all the gods, the sun was linked with gold in remote prehistory. The Greek poet Pindar called gold "the child of Zeus," who, like Jehovah, was a sun-god. As late as A.D. 1527 Calbus of Freiberg, a German physician and metallurgist, echoed the current

Over

This ancient Sumerian jewelry comes from the Middle Eastern city of Ur, located in what is now Iraq, and has been dated at circa 2500 B.C.

Facing

These gold nuggets in their various sizes and shapes are examples of those that can be found freely in nature.

opinion that "every metallic ore receives special influence from its own particular planet. Gold is the Sun or its influence, silver of the Moon, tin of Jupiter...thus gold is often called the Sun."

It was thought that metals entered the earth through rays from the celestial bodies. In a sense the ancients were on the right track. The shining nuggets they cherished are indeed of heavenly origin. Gold on our planet is truly the "gift of dying stars."

To identify gold thusly we must remember that our sun *is* a star, even if never a source of our earthly gold. The precious metal found in nature today is the by-product of stellar cyclotrons, produced from suns far older than ours. Geologists have determined from radioactive dating of rocks that the earth formed about 4-1/2 billion years ago out of a gold-and-silver- enriched interstellar medium which astrophysicists call the solar nebula. Scientists determined the sun's elements by analysis of their spectral line. They discovered the relative chemical proportions in the sun are the same as in the earth and concluded that the earth, sun and the entire solar system emerged from the same fiery star dust.

Stars and the stuff between them are composed chiefly of the two simplest atoms hydrogen and helium, both of which are light. A star is born of dust and interstellar gas. The interstellar core heats up. Nuclear reactions provide energy for the great amounts of visible light and other forms of electromagnetic radiation that pour out of the star and which, in the case of our sun, sustain life on earth.

The nuclear reactions, part of a star's life cycle, are much like those in a hydrogen bomb. The abundant hydrogen atoms burn off to form helium. Four hydrogen nuclei are converted to a single helium nucleus, and about one per cent of the initial mass of hydrogen converts into energy, according to Einstein's celebrated equation $E=mc2$.

Eventually a star burning hydrogen will consume the fuel in its core. Then it must burn the converted helium. Thus, the star evolves, burning first hydrogen, then helium; and when that is exhausted, carbon. Finally the stellar core turns to pure iron, which is inert and cannot under usual conditions be used as fuel.

Lighter stars then gradually cool off and die. However, massive stars, larger than our sun, undergo more cataclysmic explosions. The unstable stars generate elements heavier than iron. These red-hot exploding stars are supernovae. In a last great surge they become as brilliant as the whole galaxy, emitting heavy elements like gold, silver and uranium.

How did the gold created by such incredible detonations get to the streams of California, the Klondike, the gold fields of South Africa and the Egyptian desert? The exploding materials traveling at three thousand miles a second spewed out into the interstellar medium. They floated about in tenuous gases until an interstellar cloud formed. The cloud then collapsed under local gravitational condensation, and new stars and their accompanying planets formed.

Our sun and the small rock and metal-laden planets closest to it: Mercury, Mars, Earth and Venus were born when the weighty elements solidified out of clouds of cosmic gas. By the time the more distant planets formed little remained of the heavy elements, which had precipitated earlier on the dense planets closer to the sun.

As the earth cooled gold, which is extremely heavy, settled beneath the planet's mantle. It is widely distributed in a thin layer throughout the crust. Today gold exists in a variety of rock matrices. A deposit may be of any geological age and associated with sedimentary or igneous rock.

Toward the end of the Cretaceous period, perhaps 94 million years ago, the molten interior of the planet began to boil and convulse. Mighty movements flexed and snapped the earth's surface, thrusting up mountains and volcanoes, twisting and wrinkling the planet's crust. Internal pressure forced molten magma up to form the masses of rock, often granite, which form the base of mountain ranges.

Gold is most often found near granite. The most common matrix for gold is quartz, but veins or lodes are also found in calcite or limestone. The vein material generally includes other metallic compounds. Pyrite, the brassy iron sulfide which resembles gold, is abundant in some veins. Because it is easy to mistake it for the real thing...its name, "fool's gold."

In 1576 John Lok, whose family made a fortune in the West African gold trade, underwrote Martin Frobisher's voyage to find a northwest passage from England to the Orient. Frobisher, a privateer turned explorer, sailed for the southern tip of Greenland with three small ships. Only one ship returned to England. It brought a curious souvenir — an Eskimo and his kayak. Samples of black earth from Newfoundland excited even more interest. Alchemists studied it and proclaimed that it was rich in gold. Lok sent Frobisher back to get more. Queen Elizabeth herself invested £1,000 sterling in the venture and loaned Frobisher a Royal Navy vessel. He brought back hundreds of tons of the ore which turned out to be worthless iron-pyrite.

Ore deposits of many metals, including gold, silver, tungsten and copper, were created during and after the period of igneous activity. Ascending magmas bearing the metals in suspension forced their way through the earth's crust into cooler zones of subsurface rock. The metallic elements were deposited in fissures and cracks of faulted rocks as veins or replacements.

If the gold was in an active solution and was deposited in a rock such as limestone, which is unusually prone to chemical action, it could penetrate solid rock. The solution slowly dissolved the rock, replacing it with gold, atom for atom. These are replacement deposits. More often gold appears between rock fissures in veins that wind through the host rock as veins traverse the human body. Frustrated prospectors discovered, however, that these veins follow irregular routes, varying in patterns, length, thickness and depth. A vein as thin as a hair ribbon at one point and quite close to the surface can fragment into almost invisible branches, plunge deep and later emerge as a rich pocket. Veins extend into the earth in tabular form...in sheets or planes at an angle. A promising vein may be several miles long or terminate abruptly after an inch.

Gold, in veins or replacement deposits, is almost always found together with a gangue, or host material, which keeps it in place. Rock and the gold imprisoned in it make up the ore. The gold does not have to be highly visible for its ore to be worth processing. Much of South Africa's gold comes from ore in which the gold is invisible but recoverable thanks to modern refining technology. Gold, because of its stubborn refusal to form combinations, is most often found in relatively pure form in nature even though it is almost always alloyed with some silver. The gold and silver alloy *electrum* made King Midas rich and famous. Gold occasionally occurs combined with tellurium as gold tellurides and is a by-product in the refining of certain copper, lead and zinc ores. The bulk of gold is uncombined, or free, either in primary vein deposits or in secondary alluvial deposits.

The earliest gold mankind found was alluvial or placer gold, which had been freed by erosion from its subterranean prison, torn away and washed into streams and rivers. This was true of most of the gold produced before 1873, when vein mining was introduced to work the rich lodes in California, Nevada and Colorado. Through the geological ages weathering forces such as rain, wind, freezing, thaw and plant growth have worn down the earth's surface. These agents of nature have "mined" gold for eons, removing thousands of feet from mountain ranges like California's Sierra Nevada.

The products of disintegration, ranging in size from massive boulders to pebbles and even finer particles, found their way into watercourses which tumbled and abraded them. The farther chunks of ore traveled, the more the precious metal was separated from the gangue or debris. Surface water carried away some of the constituents of metal-bearing minerals that are subject to chemical decomposition during erosion. Silver is not found in alluvial deposits because it decomposes upon contact with the chlorides that occur in rain water. Gold, impervious to chemical change, retained its identity as it rushed down prehistoric streams.

The high specific gravity of gold, 19.4, assured that the sorting action of the water would carry off lighter sand and gravel debris while bits of gold dropped to the bottom. Because gold is about seven times as dense as gravel or sand, gravity helps concentrate it in sand or gravel bars and in "potholes." Where the stream gradient is reduced or moved slowly around curves, gold particles are trapped in cracks or between stones. Fragments ranging in size from powdery dust to masses of a hundred pounds lay awaiting discovery and collection.

Placer deposits are not always on the surface. In many locations, especially California and Australia, there are gold-bearing gravels buried deep beneath thick flows of ancient basaltic lava. The gold in these fossil placers was deposited in prehistoric stream beds and subsequently welded to deep rocks by geologic processes and covered by igneous or sedimentary rock. Sometimes these subterranean placers were exposed eons later when other streams channeled through the overburden of lava and sediment.

Gold is found as dust, flakes, grains or nuggets and in curious crystalline branching formations. It was these plant-like structures that supported speculation that gold grew as a plant in the rocks. "As fishes die in the water," wrote Thomas Norton a 16th-century chemist, "so metals generated in the earth are subject to rust, corruption and gradual destruction above ground." His tract was devoted to

35

a plea for the closing of gold mines. Many believed that gold deposits should be left alone for certain periods so that nature might replenish them.

In Borneo nuggets were "gods of the soil" and played an important part in propitiatory rituals. Malay miners believed gold in the earth belonged to a golden deer who chose to give or withhold it. Prayer and fasting preceded gold gathering in various parts of the East Indies where the miners were careful not to remove all the gold they saw lest the "soul" of the gold depart and no more "grow."

Gold can, in fact, be grown as well as mined, but it is hardly profitable. Horsetail, a common plant of the United States and Canada, has such an affinity for gold that it has been burned for its metallic content. A ton of horsetail grown in low grade gold fields yields as much as 4 1/2 ounces of gold. Two other plants whose juices trap fine gold are the gogo of the Philippines and the itambamba of Brazil.

Although gold is found virtually everywhere in the world, the major concentrations eventually began to have a great bearing on population distribution, especially after the discoveries made by the 16th-century Spanish in the New World.

The greatest sources of gold and the longest-worked deposits are located in the vast continent of Africa. Beginning thousands of years ago men systematically collected gold from the plateau between the Nile and the Red Sea, the sands of the Nile, Nubia in the northern Sudan and Ethiopia as well as in southeastern, central and western Africa. The incredibly rich gold fields of South Africa, however, had to await exploitation made possible by 20th-century technology. Today South Africa, which a century ago produced less than one per cent of the world's gold, accounts for over 70 per cent of the annual production, excluding the Soviet Union, which was the second largest producer before the nation fragment. Output here was and remains a state(s) secret.

Europe's gold is concentrated southeast from the British Isles through France, along the Alps and Balkan mountain ranges and in rich deposits on the Iberian peninsula. Asiatic Russia has colossal gold concentrations and considerable amounts of the metal have been produced in Asia Minor, Arabia, the Hindu Kush and southern and eastern India. Quantities have been accumulated from the Chinese coast, Korea, Japan, the Philippines and through the Malay archipelago across the Sumatra, Java, Papua, New Guinea, Australia, New Zealand and Fiji, where gold miners recently struck for a noontime "sex break."

In the Western Hemisphere gold was deposited primarily down the western coastal areas from Alaska and northwest Canada to the Rocky Mountains, along the Pacific coast of the United States into Mexico, across Central America and over the northern half of South America.

The known deposits of gold were widespread in antiquity. Gold production in ancient Egypt was so extensive that it almost constituted a monopoly for several thousand years. Gold deposits were worked in India, Africa, Persia, Arabia, the Caucasus, Asia Minor and the Balkans in the millennia before Christ, but production was neither as great nor as continuous as that of the Nile kingdom.

The very earliest discoverers of gold limited themselves to picking up random grains or nuggets which lay superficially exposed in watercourses or on the ground. The Greek legends of the Golden Age may trace their origin to the time when men first began to look for gold systematically. Gold seekers were few in number, and the gold of infinite ages lay waiting like rich fruit ripe for the harvest. The accumulation of millions of years shone in the limpid shallows of rivers and streams or appeared just below the earth's surface when heavy rain, an avalanche or wind stripped away the topsoil.

Ancient writers record instances of gold found when trees or plants were uprooted. Indeed, in modern times gold nuggets have emerged in the course of digging foundations, wells and vegetables. In the early 19th century when a great deal of gold was found in the southern Urals, many nuggets of varying size lay spread under the soil, no deeper than a crop of potatoes.

The Golden Age

Men, always longing for gold, associated it with an idealized vanished age of great happiness and prosperity. They also ascribed golden attributes in both a physical and a metaphorical sense to a paradise or abode of the blessed dead which lay somewhere beyond and to which they could aspire and attain through leading the proper kind of life. The fifth Mohammedan heaven is of gold. The biblical Jerusalem with streets of gold typifies the golden dream. The Hindu epic the Rig-Veda stipulates that he who gives gold will have a life of everlasting light and glory.

Eight hundred years before Christ the Greek poet Hesiod wrote of a happier time when the gods created man and lavished precious gold on him. The sands of the rivers were all of gold and a great river called Ocean flowed all around the earth, which was never troubled by wind or storm. On Ocean's bank was the blessed abode of the dead, where golden flowers blazed on the trees. Gold high on the mountains was liquefied by forest fires and gushed in molten streams over the earth. Whole islands were made of the precious stuff.

This was the Golden Age. The earth and the flocks were fertile and man lived in golden plenty and peace. This is how some choose to recall the lost ages when mankind made the transition from the harsh existence of hunter-gatherers to the more favorable conditions of settled life and a predictable food supply.

Hesiod wrote that the gods experimenting with the metals from the highest to the basest made men first of gold. But they disappointed their creators with their behavior, so the gods made mortals of less precious silver. However, the silver men were far less intelligent than their ancestors and warred incessantly with their brothers. The gods succeeded them with an inferior race of brass or bronze men who were even more bellicose and violent than the silver. Finally, lamented Hesiod, came the contemporary age of iron filled with toil and sorrow.

In the 5th century B.C., Plato explained the scarcity of gold in his time as a punishment by the great Zeus for man's evil ways. Once there was a time when the first men lived in harmony and golden luxury with the gods. This earthly paradise was on the lost continent of Atlantis, which lay due west of the Pillars of Hercules that guarded the entrance to the Mediterranean Sea. The golden spires of Atlantis blazed forth to light the whole world and guide Apollo as he drove the chariot of the sun across the heavens each day.

This Golden Age came to a halt when men and gods began to mingle too freely and beget such base beings that Zeus judged them undeserving of this profusion of his gold and blessings. He caused Atlantis to sink beneath the sea and ever since, Plato explained, men have had to search and toil painfully for bits of gold once so abundant.

Quality of Gold

The quality of gold today is measured in terms of karats and in parts per thousand. The word "carat" derives from the Italian *carato*, the Arabic *qirat* and the Greek *keration*, all of which refer to the fruit of the carob. This leguminous tree has hornlike pods containing seeds which were once used in Oriental bazaars to balance the scales weighing gold, gems and pearls.

The dried seeds are amazingly uniform in weight. Each weighs about a fifth of a gram, which is the weight now assigned to the metric carat. In the case of gold, the karat is usually spelled differently and is not a measure of weight, but one of purity. Pure gold is 24 karats gold. Because pure gold is so soft, it is usually alloyed with a small amount of copper, nickel, zinc, silver, or other inferior metal to increase its resilience without diluting its beauty.

Fine jewelry is seldom made with gold of a purity less than 18 karats 18/24, or 3/4 pure gold. In the U.S. gold 14 karats and under accounts for the majority of retail sales; however, people in most countries prefer purer, but softer, gold. An increasing amount of mass-produced jewelry is being made with 9-karat gold, which contains only 9 parts gold out of 24, about 37 per cent. Europeans and Latin Americans prefer 18 karat gold and in the East people prefer 22-karat (91.6 percent pure) or 24-karat (99.9 percent) gold. Indians like 22-karat gold and there were riots when the Indian government attempted to outlaw possession of it in favor of 18-karat gold in order to decrease gold consumption.

The quality of gold is also measured in parts per thousand. Most of the newly mined gold put on the market by South Africa is 995 gold, meaning that it is 99.5 per cent pure. The Soviet Government refined all its gold production to 999 quality, which is 99.9 per cent pure and can only be produced by subjecting refined gold to a further electrolytic process.

Gold isn't measured on the familiar avoirdupois scale of weights with its pounds and ounces, but rather in troy ounces, a term derived from Troyes, France, where it originated. Both systems are British, although their names are of French origin.

The troy system is used to weigh precious metals and gems. The avoirdupois system, used to weigh everything but precious metals, gems and drugs, takes its name from the French *avoir de pois*, meaning goods of weight.

The two systems of measurement are based on a very ancient unit of measure, the grain, which can be used to convert weight from one system to the other. The troy ounce contains 480 grains, while 437.5 grains constitutes the avoirdupois ounce. Thus, a troy ounce equals 1.097 avoirdupois ounces. A standard 400-troy-ounce gold bar weighs 438.8 avoirdupois ounces, or 27.4 avoirdupois pounds.

Grains were adopted as a unit of weight in ancient Mesopotamia when developing commerce demanded standard units of weight to facilitate transactions. Balances were used to weigh precious metals. Observing that grains of barley or wheat from the middle of an ear tended to be of standard weight despite the size of the ear itself, merchants and traders divided the large major unit, the shekel, into grains. The grain was the hypothetical weight of a grain of wheat. In practice the shekel might vary from 120 grains to more than 200. Multiples of the shekel were introduced — the mina, reckoned variously from 25 to 60 shekels, and later the talent, equal to 60 mines. The mina and the talent were used throughout the Middle East for thousands of years.

The United Kingdom used the grain measure until it adopted the metric system upon joining the Common Market. Today almost every country weighs gold by the gram rather than the grain or pennyweight, although the price of gold is quoted in troy ounces.

Measuring Gold

Avoirdupois is the common American system of weight measure, but it is not used for gold and other precious metals:

1 avoirdupois ounce	28.350 grams
1 avoirdupois ounce	437 1/2 grains
16 avoirdupois ounces	1 avoirdupois pound
16 avoirdupois ounces	7000 grains

Instead, the Troy weight system is used for measurement of gold and other precious metals. As noted in the table below a Troy ounce is slightly heavier than an avoirdupois ounce:

1 troy ounce	31.103 grams
24 grains	1 pennyweight
20 pennyweights	1 troy ounce
480 grains	1 troy ounce
12 ounces	1 troy pound

4 — *Greatest Potential Source...*

The Seas

T he Ancients were unaware of the greatest potential source of gold of all...the oceans and seas which cover three-fourths of the earth's surface. It has been estimated that they contain enough gold to make every human being a multimillionaire. Men have long been trying to recover gold from the sea, certainly since 1866, when a French scientist speaking before the American Association of Science said, "Gentlemen, I believe that among other things you will find gold in the ocean." Despite hundreds of theories, patented inventions and "scientific" attempts, man has thus far been unable to mine the oceans' gold profitably.

How did gold get into the oceans? It has been washed there atom by atom since earth's superheated halo of steam condensed into the seas. Some gold was deposited beneath the seabed when the planet was formed and later thrust up in the same kind of movement that shaped terrestrial mountains. Gold wrested from mountains on land and carried in streams and rivers sometimes made its way into the sea. There are rich concentrations of sea gold in areas that have been fed by the gold-bearing rivers of Alaska, Washington, California, Japan and Australia.

The first scientific estimate of the amount of gold actually present in sea water was made in 1887 by an English chemist, who figured there were 65 milligrams of gold in a metric ton of water. Estimates offered by scientists from Australia to Sweden varied widely but pointed to several interesting facts. There was more gold in solution in polar ice than in the surrounding water. The highest concentrations of all were found in plankton, the minute drifting plants and animals fed on by marine species.

While scientists studied ways to extract gold from sea water, scam artists were cashing in on various "sea gold" schemes. In 1897

41

Prescott Ford Jernigan, a Baptist minister and pillar of the community of Edgartown, Massachusetts, announced that a dream had revealed to him the method to harvest ocean gold. The process involved submerging an "accumulator, a zinc-lined wooden box filled with chemically treated mercury, a favorite ingredient for centuries in gold-swindling schemes. When an electrical current was passed through the box, gold would supposedly be absorbed by the quicksilver.

Jernigan convinced two wealthy parishioners to invest in his project. A box was built and a deep-sea diver hired to submerge and connect the device. On a cold February night the box was lowered over the side of a small boat in Narragansett Bay and left for a full running of the tide. When it was hauled up, government assayers found $5 worth of pure gold in the box — a promising start.

The parishioners and a few other men put up $20,000 for more extensive tests. In December excited investors formed the Electrolytic Marine Salts Company, capitalized in Boston at $10 million, an enormous sum at the turn of the century. Within the first year 700,000 shares were sold on the strength of 250 accumulators which allegedly yielded $1,250 each time the tide turned. The tide finally turned for Jernigan when it was discovered that he and his diver had been salting the accumulators with gold.

Many more millions were expended on the quest for sea gold by chemists, engineers and governments. In 1919 the Germans hoped to extract enough gold to pay their war debt. They asked celebrated chemist Fritz Haber to tap the inexhaustible supply of marine gold. Haber was a brilliant man who had helped his country in World War I by synthesizing ammonia so that Germany was well supplied with explosives throughout the war despite England's blockade which kept out Chilean nitrates.

Haber organized the German South Atlantic Expedition outfitting his ship *Meteor* with a complete laboratory and filtration system. For ten years he plied the oceans, testing water from the mouth of the Rhine to the South Seas. He collected samples from the California coast and the polar ice caps and scrutinized thousands of marine plants and animals.

Initially he was pleased at the amounts of gold his tests revealed. He discovered, however, that minute traces of gold in his testing chemicals themselves were falsifying the results. When the assay methods were refined, he found very little gold. He wrote a letter to a leading German chemistry journal in 1929 summarizing his decade of travail and commenting, "There is more chance of finding a needle in a haystack than extracting gold from the sea."

The Japanese conducted experiments in the early 1930s, and then the United States, spurred by the rising price of gold and certain promising results, took over. In 1935 Dr. Colin Fink, an internationally respected authority on electrochemistry, developed an electrical method to collect marine gold. Jernigan, the bunco artist, had been on the right track after all with his electric current.

Fink's plan involved a simple procedure connecting metal plates to positive and negative terminals of a battery, inserting them in the sea and letting the gold accumulate at the negative plate. He said it would be feasible for an ocean liner to "use its propellers as cathodes and plate them with gold during a voyage." Unfortunately Dr. Fink's *eureka* was a bit premature. The negative plate had to spin at a rate of 15,000 revolutions per minute for the procedure to be effective. Thus, it cost five cents to collect a penny's worth of gold.

In fact, the relatively minute amount of gold in vast bodies of water destroys the economic incentive for extracting it. On average there are only one to two parts of gold to a million parts of water, although the concentration can be higher. In 1951 water tested from the Sea of Japan showed thirteen milligrams per metric ton. Scientists proposed to extract it by evaporation using sunlight as the source of heat, but nothing came of it.

Georges Claude, a French inventor who had made a fortune from neon lamps and liquid air, set up an experimental extracting plant in 1936 aboard a steamer that made the Pacific coastal run between North and South America. His extraction plant consisted of a large funnel-shaped apparatus which forced sea water into a long iron tube filled with a compound of iron pyrites. Since finely divided pyrites attract gold, the gold in water flowing through the pipe should theoretically have remained with the pyrites. Claude found that there was scarcely one milligram per metric ton in the water. His disheartening results were confirmed by the experiments of others.

Filtration and evaporation offer two ways of collecting ocean gold, but both present problems which make their application unlikely. Massive quantities of sea water would have to be processed to yield appreciable amounts of gold. If a totally efficient process were developed, a cubic mile of sea water would have to be screened to produce fifty pounds of gold. In other words, five million tons of sea water, more than a billion gallons, are required to yield one ounce of pure gold. In a cubic mile of sea water there is about $370 million worth of gold. To treat such a volume of water in a year would necessitate filling and emptying 200 huge tanks of water, each 500 feet square and five feet deep, two times a day. Such a feat is not impossible but it is economically out of the question. Simple logistics bobble the mind!

Human nature being what it is, men will keep on trying to separate gold from water. Nuclear-powered desalinization plants which would produce not only gold but fresh water, power and salt have been proposed by several researchers who predict they will be in use by the end of the century.

Mining and mineral exploration interests have recently prospected off the east coast of Canada for undersea gold placer deposits. Scientists studied offshore sediment movement, past sea current and beach wave action as well as geologic formations to determine what areas were most likely to have gold deposits. Seismic acoustic profiling was employed to examine the ocean bottom. Prospective drill sites were chosen, taking into account such past and present features as ancient beach lines and buried stream channels. Test holes were drilled and the cores analyzed to determine the amount of gold present. Enough gold was found in a zone lying off former gold-producing land in Nova Scotia to warrant further exploration.

The coarse gold appears to have been left behind when erosion and perhaps glacial action wore away the top of a hill or mountain that had once been above sea level. But processing would be very difficult since the gold-bearing alluvium is in water more than 100 feet deep. Mining engineers are designing special suction dredges to remove the alluvium to a site on a ship where the conglomerate of mud, gravel, sand and gold would be sluiced. The gold would separate easily from the lighter tailings, which would then be dumped back in the sea.

So, gold from the sea may yet be a reality...thanks to new developments in mining technology, the same developments which have made it profitable to mine gold from low-grade ore and to rework the tailings from old and abandoned mining operations.

Part Two

Explorer's Gold

At no time in history was gold more critical in shaping the destinies of men and nations than during the period following the Spanish Conquest of the New World. During the 16th and early 17th centuries gold in unprecedented volume poured into Europe through Spain. At the dawn of the Conquest, however, the prevailing mood in Europe was one of disillusionment and pessimism. Most of the continent lay exhausted from decades of internecine strife. The specter of Islam haunted a disunited Christendom, whose Crusades had failed to regain the Holy Sepulcher of Jerusalem.

The Church had lost moral leadership and the papacy was in the corrupt hands of a Borgia. Hostile Turks occupied most of Greece, Albania and Serbia, pressing ever closer to the European heartland. These infidels blocked the logical routes of expansion to the spices of the Indies and the fabled gold of Cathay and Japan, which Marco Polo's narrative had brought to the attention of men and nations perennially hungry for the precious metal.

Gold was the object of virtually every voyage of discovery and every exploratory effort.

Portugal, alone among the nations of Europe, had access to the gold of West Africa and the treasures and spices of the Indies which were brought back to Lisbon by carracks plying the southern route to the East. So it was to the king of Portugal that a Genoese navigator and cartographer logically applied for backing to discover a new quicker route to the Indies. His theories were supported by the eminent Florentine geographer Toscanelli, but Christopher Columbus was turned down...a disappointment he was to suffer more than a few times.

Markings on this 17th-century Spanish gold bar indicate its weight and denote that taxes to the king and church have been paid.

Columbus

After years of petitioning at one court after another, Christopher Columbus, was received by the king and queen of Castile. Fifteenth-century Spain was essentially still a medieval country. It had been greatly impoverished when the Jews and Moors were expelled, taking their gold with them.

Ferdinand and Isabella were nearly bankrupt and keenly interested in Columbus' proposal that would bring gold to their kingdom. In fact, negotiations with him dealt in large part with gold, which was in short supply all over Europe. Columbus proposed sailing due west to reach the golden palaces of Japan and the golden wealth of the cities of India. The contract signed by Ferdinand and Isabella specifically mentioned gold. "Get gold," ordered King Ferdinand...humanely if possible, but at all costs "get gold."

Columbus was to retain a tenth of all revenues and precious metals derived from islands and mainlands he might discover. In addition he was ennobled, created Great Admiral of the Ocean Seas, and made perpetual viceroy and governor of the lands he claimed for the Spanish crown.

Queen Isabella donated her golden jewelry to launch the expedition. The rich Pinzón brothers, merchants of Palos, underwrote an eighth of the venture in return for a percentage of the profits.

This golden object dredged from Colombia's Lake Guatavita comes from a period several centuries prior to the first voyage of Columbus.

Columbus set sail with a heavily annotated copy of Polo's travels, a letter addressed to the Grand Khan of China from Ferdinand and Isabella. Visions of golden palaces danced before him.

His first landfall in the Bahamas was disappointing. The natives went about naked and lived in rude huts instead of glittering palaces. But he did note that they wore simple gold trinkets and gold nose ornaments. "I was attentive and took trouble to ascertain if there was gold," he wrote in his journal of October 13, 1492, the day after landing.

This voyage and the three that followed were, in effect, gold prospecting expeditions on which exploration and discovery were of secondary importance. Columbus was sure that he had reached the Indies, despite the absence of gilded pleasure domes and silk-brocaded potentates. He wrote a glowing letter to Pope Alexander VI, "The Island of Hispaniola is Tarshish, Ophir and Cipango. In my second voyage I have discovered 140 islands and 333 miles of the continent of Asia."

On his third voyage Columbus planned to sail south near the equator where ancient tradition asserted all precious things were to be found. He subscribed to the views of a renowned Renaissance lapidary that a sure sign of the presence of gold in an area would be the dark color of the inhabitants. Judging from the color of the natives and "from the great heat which I suffer," he wrote in his journal while sailing along the coast of Cuba, "the country must be rich in gold."

First Gold Rush

The hunger for gold that drove Columbus across the great uncharted Ocean Sea to discover the New World touched off the first gold rush in modern history. Mankind's greatest geographical discovery resulted from a monumental geographical error. Initially it paid off handsomely for Spain, a nation impoverished from the long struggle against the Moors. The first small amounts of gold brought from the Antilles whetted Europe's appetite, attracting increasing numbers of rough and ready men, many of them veterans of the Moorish wars or campaigns in Italy...men who had little to lose by embarking on a perilous voyage to the West Indies.

Gold came too late for King Ferdinand. When he died in 1516 the gold found in the Antilles had proved inadequate to alleviate the state's financial embarrassment. "Ferdinand the Catholic," wrote a chronicler a few days after the Kings death, "was so impoverished that it was difficult to procure money to furnish decent clothing for the servants at his funeral."

A golden deluge was soon to begin. Spanish adventurers made colossal fortunes, and the Spanish crown was gratifyingly enriched in the early period following the Conquest. But after the great silver mines of Potos were opened in 1545 little of the gold and silver that sailed up the Guadalquivir River to Seville remained in Spain. Most of it went immediately to pay the crushing debts owed the great banking houses like the Fuggers and the Welsers.

New World gold caused such an increase in supplies that the metal's value actually depreciated, sparking horrendous inflation felt not only in non-industrial Spain but all over Europe. Gold flowed into Spain and right out again to pay for the import of manufactured goods. Gold and silver sapped Spain's strength and emptied Iberia of much of her able-bodied male population. What gold and silver did accrue in Spain allowed her kings to embroil themselves disastrously in European politics. Precious metal financed the costly Catholic-Protestant wars and fostered extravagance and decadence. Had it been used wisely to develop domestic manufacturing and a viable economy the history of Spain and the Americas might well have been different.

Columbus returned to the Spanish court from his first voyage with samples of the gold fishhooks, nuggets and ornaments he had collected from the gentle and hospitable Caribbean Indians. As every schoolchild knows, their good will and generosity were to be rewarded with merciless exploitation, enslavement and eventual extermination in the relentless European pursuit of ever more gold.

With the conquest of the two great civilizations of America — the Aztecs and then the Inca — the floodgates opened. A deluge of gold flowed into Europe which had reached a peak of domestic production in the 14th and 15th centuries and was again starved for precious metal. For the next 250 years the New World was the world's major source of gold. By 1550 the stocks of European gold and silver had increased by half as much again as had existed when Columbus set sail in 1492. By the mid-16th century Spain had laid claim to all of Central and South America with the exceptions of Brazil and Guyana, which fell to the Portuguese by virtue of the Treaty of Tordesillas. This was the treaty by which Pope Alexander VI divided the world between Spain and Portugal. It led to centuries of strife between Catholic and Protestant European nations which used pirates and privateers to further their objectives at sea and on land.

Post-Conquest Europe erupted in a sunburst of golden splendor. Much of the American gold found its way northward across Europe into the hands of merchant bankers who gradually gained control of vast gold supplies. American treasure financed gold currencies in almost every nation. England became the most important gold center. In 1717 Sir Isaac Newton established the modern world's first real gold standard there.

A new wave of commercial expansion and prosperity followed the Conquest, and the eternal metal appeared more than ever in the pomp and ceremony of courts and churches. Dress and jewelry pleased the eye with gold worked in every conceivable manner. Burghers and merchants, even the small man, could aspire to accessories, snuff boxes and watches crafted from American gold.

The effects of such colossal amounts of gold, supplemented by production from Japan, the central European mines and later the mines of Brazil, were felt on many levels. The glorious epoch of the Renaissance was nourished by American gold. It was not only the goldsmith with his increased supplies of the malleable, gleaming metal who enjoyed the fruits of the mines. Painters and poets, all men of talent, were amply rewarded in gold by patrons moved by the Renaissance spirit. The rivers of newly mined gold infused into the European economy particularly aided the rise to power of manufacturing states, such as England, Germany and the Low Countries.

Ironically the nation that reaped the most bitter fruit from the golden harvest was Spain itself. In 1608 an Englishman noted that "whence it seems not without reason, the Spaniards say in discourse, that it [the New World treasure] worketh the same effect upon them that a shower of rain cloth upon the tops and covering of houses which falling thereupon, cloth all at last descend below to the ground leaving no benefit to those that first received it."

Columbus found no Cathay...no Japan, but like gold seekers throughout history his hopes never dimmed for long. His objective was to find inexhaustible gold mines. But at first he contented himself with the worked gold of the Indians, who gladly traded it for fishhooks, red cloth, beads and tinkling bells. When the

Spaniards asked the source of the gold, the Indians invariably pointed, indicating it lay just beyond. Some accurately waved in the direction of Hispaniola, the island now divided into Haiti and the Dominican Republic. Others maintained that much gold was found on an island called Babeque, where it was gathered in moonlight and hammered into little ingots.

On Cuba, which Columbus had initially decided was Japan, they found no gold at all. The emissaries he sent inland had returned from their mission chagrined. They had been led not to the anticipated golden city and its Emperor "el Gran Kan," but to a palm-thatched village where they were warmly received but found no court of Oriental splendor. On the trip back to the ships the men encountered what was to prove in the long run something more valuable than gold — tobacco, in the form of a cigar shared by Indians.

On to Hispaniola

Columbus sailed on to Hispaniola where he was sure his golden dreams had been fulfilled when gold-embellished Indians gathered on the shore. He invited a young chief bedecked with solid gold ornaments onto his ship. After entertaining the Indian, he had him courteously piped ashore with all honors and then noted in his journal with the callousness that marked so much of the Age of Discovery that the natives were ripe for exploitation. To the King he wrote repeatedly of how sweet the natives were, how freely they gave their gold and offered gifts of live parrots, fruits and vegetables. Of the people he had no scruples of enslaving he wrote, "They are a loving, un-covetous people, so docile in all things, I assure your Highness, I believe in all the world there is not a better people or better country; they love their neighbors as themselves and they have the sweetest way of speaking in all the world and always with a smile."

Columbus sent shiploads of these gentle people to be sold as slaves in Spain. In 1500 Queen Isabella prohibited the enslavement of the Indians; however, there were so many exceptions to this law that the Conquistadors found ample reasons for the continual enslavement of the Indians needed to work the New World gold deposits. Before long the gentle people had been wiped out as a result of the barbarous treatment they received at the hands of the Spaniards and exposure to European diseases.

Native informants spoke of Cibao, or central Hispaniola, where there were great quantities of gold. Columbus wrote, "Our Lord in his goodness guide me that I may find this gold, I mean their mines...." He fully expected to find rich mines, unaware that all of the gold had been collected from placer deposits. But before he could set off in search of the mines, disaster struck and *Santa Maria* was wrecked on a reef. The cacique Guacanagari offered condolence to the Spaniards, offering them everything of value he possessed. The villagers traded nuggets for lance points and Guacanagari gave Columbus four pieces of gold as big as his palm assuring him that he could give him a great deal more. Columbus was greatly encouraged. He decided to construct a fort ashore and sail for Spain with the glad tidings. He left a few men behind at Fort Navidad, expecting them to accumulate great amounts of gold before his return, not taking into account the fact that the gold he had been shown was the accumulation of generations of fairly casual production.

In bankrupt Spain Columbus was received with great celebration. His tales of rich gold mines, of rivers shining with the stuff, and his display of captive Indians wearing little but gold ornaments made him an overnight celebrity and riveted attention on the faraway Indies. He could enlist barely enough men to sail in 1492, but the Admiral of the Ocean Seas embarked again in September of 1493 with 17 ships instead of three and 1,500 crewmen in addition to almost a hundred stowaways, all anticipating piles of gold at Fort Navidad. Instead they found that the 40 volunteers who had eagerly stayed behind to seek the gold mines of Cibao were dead and the fort destroyed. The Spanish ruffians had roamed the island demanding gold and raping women until an enraged cacique put a stop to their depredations.

The Spaniards erected a new fort and went into the interior in search of the mines. What they found in the foothills were rivers that flowed over sands spangled with gold dust. Small grains of gold flecked the earth and they found some nuggets but no rich mines. Columbus was still loathe to relinquish his illusion of having found the fabled Indies. He was so cheered at what his men found that he let his imagination dominate good sense. In one place, concave depressions in the ground led him to conjecture that this was the very Ophir where King Solomon's fleet had found gold to build the celebrated temple.

Columbus reasoned that the rich veins which had given birth to the small grains of gold in the rivers must be high in the mountains. He sent an armed military party with trumpeters and waving banners into the mountains to seek out the mother lode. The reconnaissance party received packets of food and gifts of gold dust from friendly Indians whose villages they passed through and they also found some gold on their own but no mines. "On that trip," wrote Cuneo, a member of the party, "we spent 29 days with terrible weather, bad food and worse drink; nevertheless out of covetousness for that gold, we all kept strong and lusty." One river, the Yaqui, still yields gold to women who work placer deposits and collect gold dust in turkey quills. It was especially promising. Columbus reported gold grains the size of lentils adhered to the barrel hoops when water casks were filled.

The Spaniards watched the Indians working the placer deposits of the mountain streams of Cibao. Their method was most unsophisticated, involving no more than sorting a handful of sand for flecks of gold. The Indians gladly traded the gold for trinkets, and the party eventually returned to the main camp with about two hundred ounces of gold dust and some nuggets, one weighing 20 ounces. Indians told them of another area in the mountains where nuggets weighing more than 25 pounds were frequently found. But this story proved like so many treasure tales. The Spaniards were never able to find the spot and verify the tantalizing account. A few years later a huge nugget weighing in excess of 330 pounds was discovered in that area but unfortunately sank on a ship that encountered a hurricane while sailing to Spain.

The modest amount of gold brought back by the first reconnaissance party was a disappointment to the rough band of gold hunters who had come out to the Indies. Living conditions in the base camp left a great deal to be desired. Food was scarce because no one wanted to cultivate or hunt. There was little water and an abundance of malarial mosquitoes. The men were angry at being required to work on the construction of the settlement, to cut coral blocks for a church and to hack down trees for huts. When they refused to perform manual labor, Columbus withheld their rations. "All of us made merry," wrote one colonist, "not caring any more about spicery but only for this blessed gold."

Columbus sent samples to his royal patrons including a reported

30,000 ducats worth of gold, as well as cotton, parrots and 26 Indians from different islands. So certain was Columbus of the promise of Cibao that he requested that the Castillian monarchs recruit and send to him a contingent of skilled miners from Estremadura, Spain's chief mining center.

It soon became apparent that the gold of Cibao could only be gathered by slave labor. A poll tax in gold dust was levied on the natives of Hispaniola. Almost immediately the inheritance of generations was exhausted. The poor Indians couldn't begin to meet the quotas no matter how feverishly they washed the rivers. Failure to render the full amount was punishable by death. Every Indian over the age of 14 had to render four hawk's bills full of gold dust annually, and the caciques were committed to paying much more on a monthly basis.

Bartolomé de Las Casas, advocate and friend of the Indians, wrote the remarkable *Historia de las Indias* in which he narrates the effect of the gold dust on the Castillians and their slaves. He indicted the abominable and irrational tribute system. Labor as they might, the Indians were unable to produce sufficient gold by sluicing streams or felling trees to clear the land which they washed. Even after the quotas had been halved they could not be met. Armed Spaniards rode into the hills with their dogs hunting down and slaughtering those who tried to escape the gold workings. Great numbers died at the hands of their torturers; others starved to death in the mountains or took cassava poison.

By 1508 a census listed only 60,000 remaining natives of an estimated 1492 population of 300,000. In 1519 only 2,000 of the defenseless Indians remained. Slaves were already being imported from Africa to work the mines. By 1548 Fernández de Oviedo reckoned no more than 500 were still alive.

In 1496 the intolerable tribute system was transformed into a new form of exploitation. Under the *encomienda* or *repar-timiento* system an individual was granted lordship over a certain number of Indians. He was, in theory, to protect them (one wonders from what) and instruct them in Christianity. In return they would gratefully labor for their masters. It was in practice a much abused system of vassalage derived from the feudal estates of Spain.

Columbus, an Italian, had enemies at the Spanish court as well as among the disenchanted gold seekers who accompanied him.

Detractors claimed that he had found very little gold on Hispaniola and had salted his meager supply with "Guinea gold" from West Africa to bait the hook for investors. Some said that he had "made" the gold. A Seville goldsmith who had been with Columbus told the court he had assayed the gold from the Indies and none of it was genuine.

Columbus' grandiose plans for Hispaniola envisioned a permanent settlement of 2,000 inhabitants barred to foreigners, Jews, infidels and heretics where a closed season on gold hunting would ensure the cultivation of crops. The colonists would be granted a gold-gathering license in return for manual labor. He recognized that it would be difficult to get Indians to come to the Europeans. Each Spaniard would go to the interior to trade for gold or get it however he might and then return to the town at stated intervals to hand over his gold for smelting and removal of a percentage for the crown, for Columbus and for the Church.

In 1495 master miner Pedro Belvio was sent to Hispaniola by the crown. He brought supplies of mercury from Spain and men from Estremadura who were trained in the amalgamation process which separates gold from ore. Spain sent more colonists to erect a mint on Hispaniola, the first in the New World.

Even as Spain's empty coffers began to fill and the faltering nation revive, Columbus, a proud and sensitive man and a foreigner, saw his star fall. His many enemies prevailed, and reports of mismanagement prompted the royal couple to order him arrested and brought home in chains. When the Admiral of the Ocean Seas arrived in Spain, he was unfettered and the charges against him dismissed. He passed the remaining few years of his life an embittered man, however, stripped of titles, honors and pecuniary rights. Broken in health and spirit, Columbus made vain and humiliating appeals to Isabella, who was on her deathbed, and then to Ferdinand.

The man who once had written "Gold is most excellent. With gold the possessor of it does all that he desires in the world and arrives at sending souls into paradises [by the paying of masses]" bitterly watched the heavy laden ships unloading great chests of gold from the Indies "and none for me." It was to beckon to countless thousands in the following centuries. The admiral's "Indies" yielded more gold than mankind had ever accumulated in history. Ironically, those who craved it were rewarded with sorrow far more often than joy.

Columbus' New World ventures left him bitter but not poor. He had acquired a great number of gold ornaments in Veragua (Panama) on his fourth and last voyage and had some of the revenue due him from the diggings on Hispaniola. He had been given a chest of gold coins by Ovando, who replaced him as governor on Hispaniola, and was permitted by Genoese bankers in Seville to draw on 60,000 pesos d'oro he claimed to have left on the island. But he felt that this was far less than his rightful share and made persistent attempts to get more. Thus, all the gold he gained brought him very little pleasure.

Nicolás de Ovando, who replaced Columbus as governor of Hispaniola in 1502, arrived at the island with 2,500 men. The handful of men they found in the camp said that a native woman had recently found a huge nugget and everyone had gone to find more. Fired by such exciting news, the inexperienced Spaniards disappeared into the jungle interior where more than a thousand soon perished from disease and starvation. Those who survived often found that when the official melting came every eight months they had no gold left by the time the King's fifth was taken and they had paid back the amount advanced over the season for provisions. As in the 19th century North American gold rushes it was the shipowner and shrewd merchant who grew fat while the miner, pursuing a golden will-o'-the wisp, remained poor.

Columbus had found a promise of gold on the South American mainland. On his third voyage he touched on the lush coast of Venezuela. Naked Indians adorned by large polished disks of gold around their necks welcomed the Spanish expedition. They were delighted to barter their gold, preferring objects of copper whose odor they sniffed with pleasure. They deemed copper, which came to them only through trade with Central America, far more valuable than gold. The Venezuelans were more advanced in metallurgy than the inhabitants of the Caribbean Islands. They made objects of guanín, called today tumbaga, which is an alloy of gold, silver and copper in which the gold content ranged from 9 to 80 per cent and the copper from 11 to 74 per cent. Their sophisticated goldsmiths had discovered that when the copper content of a gold alloy was between 14 and 40 per cent the melting point was reduced from the 1,073 degrees Fahrenheit of pure gold, making the metal more easily workable.

6 — *Cortés, Greedy Adventurer...*

Conquistadors

I n the great age of discovery the most stunning achievements and the most incredible barbarities were the products of the determined search for gold by men who let nothing stand in their way. The chimera of gold lured Magellan to his death in the Philippines, Balboa to the Pacific, Cortés from Mexico to Lower California, De Soto from Nicaragua to Peru, Coronado throughout the Southwestern states. Gold lust compelled countless others to explore the most forbidding zones of the Western Hemisphere. These men were the fabled conquistadors.

By 1520 most of the large islands of the West Indies had been explored and a few settled. Cuba served as the base from which expeditions fanned out hunting for precious metal. There was as yet little effort to establish settled colonies with an agricultural base. The adventurers who came out of Spain were the restless, reckless and greedy men who have always flocked to gold fields. At the rumor of a new strike they would drop everything. Heedless of any hardships lying ahead, they plunged into labyrinthine jungles and trackless deserts.

Arrogant and contemptuous of danger, the conquistadors were drawn to America by the promise of easy riches, glory and honor. They came from a poor land, from poor families or impoverished noble families. They had been raised on the chivalrous Christian romances of the Middle Ages. They knew the way was hard with death facing them at every step; however, if they survived they could return to Castile as rich men.

In 1513 a gold hunter named Vasco Nuñez de Balboa went prospecting on the mainland with a small group of Spaniards. He crossed the Isthmus of Darién and became the first European to see the Pacific Ocean. At the time it seemed far more significant that he had found an abundance of gold. His men pillaged every Indian settlement they came upon nestled in the verdant mountain valleys of the Isthmus, torturing Indians and throwing captive chiefs to the dogs. One of the adventurers in the band was stirred by the tales he heard of a land to the south where gold was so plentiful it could be caught by fishing with nets. He was Francisco Pizarro, who later found that southern land Peru and made the most stupendous gold haul in history.

The Spanish king was excited by Balboa's discovery of auriferous gravels. He sent a large expedition of 1500 men to Panama. Among them were priests, a governor and a *veedor*, an official overseer who accompanied every expedition to assure that the King got his royal cut. The priests were responsible for the absurd "requerimiento," a notification the invaders were required to read to the Indians before making war on them. It was inevitably "read to the trees" and heard only by assembled Spaniards, who approvingly listened to the terms which claimed all the lands, gold and Indians as vassals and property of the Pope and King.

Gradually the Spanish conquistadors ranged farther south and inland from both coasts, finding even richer sources of gold. Following streams inland up to the mountains, the prospectors often found the "purse" or rich mother lode from which *pepitas*, little nuggets, were torn and tumbled in streams and rivers. Colombia was by far the richest area for placer deposits. Gold was almost everywhere. By the middle of the 18th century this vast area became Spain's main source of gold. In the mid-19th century there was still enough to keep 15,000 persons employed in the workings.

The gold the conquistadors found in South America would have dazzled Columbus. Panama, Venezuela and Guyana, although not as rich as Mexico, Peru or Colombia, had ample supplies of gold. Chile and Ecuador on the Pacific coast were also exploited for tremendous amounts of gold eroded from the rocks of the towering Andes, source of all the Inca gold. From 1492 until 1600 it has been roughly figured that Spain's New World workings yielded a minimum 500,000 pounds of gold. The actual amount is undoubtedly higher since much gold was never submitted to officials for assay and taxation. Some of it stayed in the colonies, some was clandestinely traded to foreign merchants whose ships violated Spanish prohibitions to trade with the colonists and a vast amount, as shown by recent discoveries of sunken ships of the period, was smuggled back to Europe. Spanish shipwrecks have been excavated recently which yielded twice and even three times the gold that was listed on the official manifest.

In the 16th century the greatest mother lode of them all was the Andes chain. The great mountain system, which stretches 4,500 miles from Panama to Tierra del Fuego, is still a major producer of gold, platinum, silver, copper, tin, lead and other minerals. After Balboa's sighting of the Pacific the Spaniards established themselves in Panama and began the penetration down the west coast of South America into Ecuador, Bolivia, Peru and Chile. The abundance and purity of the easily collected placer and alluvial gold deposits they found astounded even seasoned prospectors. For hundreds of years after the great mass of Inca treasure had been seized, the volcanic mountains continued to fulfill their early promise.

The treasures of the Incas, plus the Aztecs to the north in Mexico far exceeded any visions ever stimulated by Marco Polo or the story of Solomon. The speed with which the conquistadors led by Pizarro and Cortés overwhelmed the two great centers of high American civilization was astonishing. Spaniards were small in numbers, their horses few, the climate and terrain hostile; yet Cortés in Mexico and Pizarro in Peru felled the two great empires with almost lightning rapidity.

Cortés in Mexico

Mexico had been discovered by Juan de Grijalva less than 10 years after Columbus first set foot on the continent. In 1519 the governor of Cuba sent out an expedition to conquer Mexico. At the head of the force of about 600 men and 16 horses and equipped with only a few small cannons was a brilliant, shrewd adventurer named Hernán Cortés. A dapper man who habitually wore a diamond ring and a massive gold chain, Cortés had a Cuban land grant or *encomienda* worked by Indian vassals. "How many of whom died in extracting this gold for him, God will have kept a better record than I," commented Las Casas.

Cortés left Cuba despite a last-minute order from the governor, who feared his ambition, barring his departure. After first making a brief stop on the island of Cozumel, Cortés landed near Vera Cruz, dismantling his ships to keep his men from deserting, and founded a town in an attempt to legalize his independent command. The

Spaniards headed through the steamy, jungle-covered mountains toward the Aztec capital accompanied by 1,000 cannibal Cempoalan Indians eager to fight their Aztec adversaries. The Spaniards were aided in their conquest, both in Mexico and in Peru, by exploiting the lack of unity between various native groups. In addition they were able to make use of local legends and weaken enemy resistance by playing on superstition.

In Yucatán hospitable Mayan chiefs had given Cortés gifts of gold, food and women. Among the women was an Aztec princess who became his mistress and interpreter. Doña Marina, as she was called, had been given as a child to a Mayan chieftain. She proved invaluable to the Spaniards when they reached the Aztec capital.

Messengers from Montezuma, the Aztec emperor, begged "the white gods" to stay to the east and proceed no farther. This emissary brought food and rich gifts of intricate cast-gold jewelry. Once the Spaniards had seen the glint of gold there was no question of turning back as Montezuma asked.

A Mighty People

The Aztecs were a mighty people with an all-powerful bloodthirsty religion. In the 12th century they had achieved political ascendancy over the Toltecs in the fertile area of central Mexico. By the 14th century they had founded the capital city of Tenochtitlán, a New World Venice on a small island in Lake Texcoco. Gradually the warlike Aztecs subjected the people of the surrounding areas.

By the time of the Spanish Conquest they had established a large empire with a complex military, religious and civil bureaucracy. Vassal states rendered tribute in gold and provided sacrificial victims essential to their formidable religion. The sun-god, who died every evening, required copious amounts of fresh human blood, the precious liquid of life, to assure his daily rebirth at dawn.

According to one legend, the Aztecs had conquered the Itzá, whose great god was a tall bearded white man. He was known as Quetzalcoatl to the Itzá and Kulkulcán to the Maya. The Aztecs included him in their pantheon of gods. They erected the highest temple pyramids in his honor and lavished gold and human blood on his altars, for they believed the god had sailed away to the rising sun, promising the Itzá he would return and avenge their enemies.

Montezuma was uncertain and puzzled by the arrival of the Spaniards. Perhaps the bearded strangers were the fulfillment of the prophecy? Certainly, they must be divine beings. Accordingly, he sent offerings worthy of the gods.

Aztec informants later reported to the chronicler Bernardino de Sahagún, "They gave the 'gods' golden necklaces and ensigns of gold and quetzal feathers. And when they were given these presents, the Spaniards burst into smiles; their eyes shone with pleasure; they were delighted with them. They picked up the gold and fingered it like monkeys; they seemed to be transported by joy, as if their hearts were illumined and made new. The truth is that they longed and lusted for gold. Their bodies swelled with greed and their hunger was ravenous; they hungered like pigs for the gold. They snatched at the golden ensigns, waved them from side to side and examined every inch of them."

Montezuma's bewilderment was turned to Spanish advantage. Cortés sent him presents and a message that he had come from the mightiest King on earth who desired to establish trade relations with him. The conquistador sent a Spanish helmet back to the capital asking that it be brought back filled with grains of gold so he might see what their gold looked like. "Let him send it to me," said Cortés, "for we Spanish have a disease of the heart which can only be cured by gold."

The helmet was returned brimming with gold dust, "just as they got it from the mines." The Emperor sent other gifts; the most splendid was a huge sun disk inscribed with the signs of the Aztec calendar. It was as big as a cart wheel and weighed almost two hundred pounds and was made of fine gold. There was an even larger wheel of silver, many cast-gold ducks, a large number of charming animal figures in gold, jewels and exquisite colorful featherwork. With these items Montezuma sent yet another request that the strangers proceed no further.

But now the gold fever burned too hot. Cortés marched and fought for three months through searing heat and numbing cold, across mountains, rivers and jungles. Montezuma sent frequent emissaries with gifts and pleas to the white gods to turn back, but they pushed relentlessly on toward the setting sun. Then, one day they looked down through the crystalline air to the gleaming canal-laced city of 300,000 which lay 7,500 feet below in the Valley of Mexico. As they descended to the great causeway that led to the capital, Montezuma came out to meet them. He was no half-clad savage but a golden-crowned monarch, borne on a shining gilded litter and clad in superbly colored garments of cotton, embellished with gold and jewels plus a mantle of iridescent feathers. The very soles of his sandals were of gold.

Bowing to the inevitable, the emperor graciously welcomed the bearded gods. They were lodged in the palace that had belonged to his father and given rich robes, women and slaves who served them from gold platters. The unrefined Spaniards were stunned at such magnificence and hospitality. The Aztec leader offered them many splendid gifts. In spite of all this, the men realized they were in effect prisoners in a gilded cage. Exit from the city was controlled by drawbridges and the Spaniards were greatly outnumbered. The frequent agonized screams they heard from sacrificial victims at a nearby temple did nothing to allay their growing anxiety.

While exploring the large palace, some of the men had discovered a secret treasure room where the wealth accumulated by Montezuma's father was stored. The sight of great slabs of refined gold, masses of finely wrought gold jewelry, plate and ornaments along with treasures of gems and silver intoxicated the men. They were no longer content to stay confined as honored guests in the palace. With a boldness born of desperation Cortés seized the emperor and kept him chained as the Aztec tribute collectors brought mounting piles of gold to the Spaniards. The men were unbelieving at what they saw; some wondered whether they might not be dreaming it all. The Aztec emperor was pressured to assign

Over

Author Jenifer Marx dives for conquistador treasure on a wreck in the Bahamas where she found a gold coin, below, and a scabbard tip.

Facing

This 12th-century Aztec pectoral medallion is typical of the golden ornaments plundered by the conquistadors of Cortéz when they overran Montezuma.

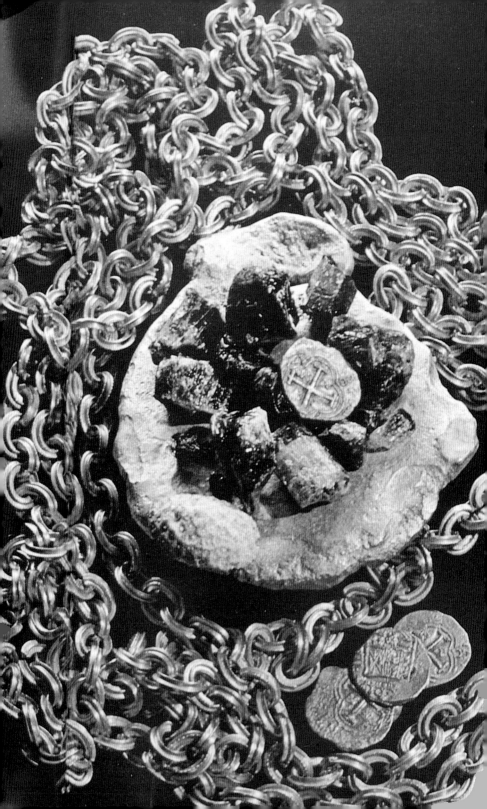

men to lead Cortés' soldiers to the mines in Cuzula, Tamazulapa, Malinaltepic, Teniz and Muchitepeque. The soldiers returned with glowing reports of resources rivaling those of Solomon.

Mexico was exceedingly rich in gold; but because the Valley of Mexico had no native gold, the Aztecs had organized several methods of acquiring it. Some they got in trade but most came in tribute from subject towns. Cortés queried Montezuma at length and was pleased to learn that nearly two tons of gold a year reached the capital in various forms. The Tlapanec's tribute, for example, consisted of 20 chocolate jugs full of gold dust and ten gold plates each four fingers wide, two feet long and the thickness of parchment. From the rich Mixtec region, which still produces some 200,000 fine ounces of gold annually, came worked gold including breast plates, collars, pendants, bells, headdresses and labrets (lip plugs) of coral and rock crystal mounted on gold.

Throughout the winter and spring of 1520 the pile of gold mounted in Tenochtitlán. As rapidly as it came the barely literate Cortés and his men melted it down. An estimated two thousand pounds of Aztec goldwork went into specially constructed furnaces to be cast into small stamped bars. Some of the soldiers had their shares fashioned into great gold chains many yards in length. These were easily transported and, like all gold jewelry, were not subject to tax.

Cortés had left Cuba against the orders of the governor, who feared with ample justification that he would try to carve out a personal empire. When the governor learned that Cortés had sent a shipment of the royal fifth to King Charles V by which he hoped to legitimize his enterprise, the governor took action. He sent a force of 1,000 men to Mexico, led by Panfilo de Narváez, who was empowered to

Over

These golden Spanish 2-escudo coins from the mint located in Bogata were the first Spanish coins made in the New World

Facing

Note the excellent workmanship on this portion of a gold belt which was discovered on a 17th-century Spanish shipwreck in the Bahamas.

deprive Cortés of his command. Leaving a garrison of only 80 Spaniards in Tenochtitlán, Cortés surprised the force near the coast, made Narváez his prisoner and easily enlisted the men on his side with the promise of gold. And, the treasure ship Cortés dispatched to the King never arrived in Spain. It was captured by French pirates, with the gold going to the French court of Francis I.

When Cortés returned to Tenochtitlán he found the garrison besieged by Aztec forces. Montezuma was stoned to death by his own contemptuous people when he appealed to them to cease their attacks. The Spaniards were overwhelmed by seemingly endless waves of Indians. Escape from the city under cover of darkness was planned. A portable bridge to breach the gaps in the causeway was secretly constructed.

The treasure was divided up and on the dark rainy night of June 30, 1520, remembered as "La Noche Triste," the Spaniards fled. They were attacked by swarms of Aztecs on foot and in canoes. The King's fifth of the treasure, laden on horses who were quickly wounded and killed, sank quickly beneath the lake. Many men flung their treasure into the water and scrambled over piled bodies of their slain comrades which spanned one of the causeway gaps. Others, reluctant to part with their gold, sank with it or were captured and later sacrificed. Most of Montezuma's treasure was lost in the bloodied lake.

A battered remnant of a few hundred men, including Cortés and his prisoner Narváez, retreated for six days. On July 7 they won an almost miraculous victory against an Aztec pursuit force at the battle of Otumba. Cortés and his hungry, poorly equipped men fought with desperate courage to rout a force reckoned at 200,000 Indians led by a prince resplendent in gold war dress. Cortés' Tlaxcalan allies helped the Spaniards consolidate their control of Mexico. But it wasn't until a year later that they captured Tenochtitlán, marking the fall of the Aztec Empire.

The Aztec Empire was dead and Charles V had gained one of Spain's richest possessions. Cortés was named Governor and Captain General of New Spain by the grateful king. He established a large plantation and devoted himself to mining, but enemies continued to thwart him and he died frustrated and bitter, like so many other gold-hungry conquistadors whose ambitions were destined to outrun their fortunes.

7 — *Looting the Golden Empire...*

The Incas

The Conquistadors had reason to believe in golden treasures. One of their number discovered the greatest aggregation of them in the world...such a horde whose wealth surpassed even their wildest dreams. Cuzco, the Inca capital which fell to the illiterate conquistador Francisco Pizarro and 200 soldiers in 1531, was the seat of the Inca Empire. It was the richest city in the hemisphere at the time, containing vast amounts of gold accumulated over two millennia. When Pizarro's men arrived, its inhabitants were still producing prodigious amounts of the precious metal under highly organized state control.

During his sojourn with Balboa in Panama, Pizarro had heard of the great southern country with an endless supply of gold. Although he had made two earlier landings on the coast of Peru and received small offerings of Inca goldwork which whetted his appetite, he was totally unprepared for the colossal amount of Inca gold in Cuzco.

The highland Incas reigned over an empire of six million inhabitants extending from the southern border of Colombia halfway down the coast of Chile a distance of 3,000 miles.

The empire was united by the official Quechua language, sun worship and a remarkable network of roads with relay stations over which runners traveled as much as 150 miles a day carrying messages and even live fish from the Pacific.

The traditional list of twelve Inca rulers begins around A.D. 1250, but the empire had reached its zenith only a century before the Conquest. The Inca treasuries held vast amounts of tribal gold. Tribute to the Great Inca was paid by every man. The very poorest made a token payment in rice.

The trustworthy chronicler Cieza de León wrote that in the final years of the empire the amount of gold extracted annually amounted to more than 381,000 pounds...over 190 tons. The Incas regarded gold as "the sweat of the divine sun." Gold nuggets were the sun's "tears." Every aspect of its production and crafting was carefully controlled by the state. Raw metal was doled out to the *kori-camayoc* ..."he who is in charge of gold."

Goldsmiths fashioned splendid, dramatic and highly sophisticated pieces. Commoners were forbidden to keep or use raw gold or ornaments. The collecting of metals was an obligation of the great mass of commoners who labored for the Great Inca and his nobles. The Peruvians regarded the mines as living spirits and worshiped the hills and gold-bearing rivers.

All Inca gold was alluvial. It was carried as dust, spangles, grains and nuggets in the high rushing streams of the altiplano and came especially from deposits in the Valley of Curimayo, northwest of Cajamarca. But above alluvial gold there is always the vein system from which it has been torn away. Eventually the Spanish found such lodes in the formidable crags of the Andes and organized mining operations.

As early as 800 B.C. coastal Indians panned placer deposits of gold in Peru. The Chavíns, Peru's earliest goldworkers, crafted many objects of both decorative and utilitarian nature. More than a thousand years before Pizarro they had discovered that soft gold nuggets hammered into sheets eventually became brittle and broke under the repeated blows but could be made malleable again by heating the metal. Chavín artisans manufactured repoussé pectorals, ear spoons, flutes, plaques to be attached to garments, crowns and even tweezers of gold. They made a great advance when they learned to solder individual pieces of gold together to form large flexible pieces and figurines.

Goldsmiths made use of large sheets of polished gold for dramatic effect and crafted sculptural pieces of great beauty characterized by the play of flowing lines. Mochica smiths combined shell, turquoise and other materials with gold in elegant mosaic motifs. In Ecuador goldsmiths learned the secrets of cire-perdue casting and of fusion gilding of copper — alloying gold with silver and then copper.

The Incas believed in resurrection after death. Each dead Inca leader was provided with an imperial palace which was maintained by a complete staff and furnished with precious appointments made of the sun's metal. The mummified body of the Great Inca was kept with his predecessors in the great Temple of the Sun. A life-sized golden effigy, his *pucarina*, was cared for as if it were the deceased ruler himself. Chronicles tell how the retinue of one pucarina bore him in a golden litter to the palace of another dead ruler for "visits."

During his lifetime each Great Inca prepared for death by constructing his palace, covering its walls with great plates of gold and silver and filling it with treasure. In the palace of Huayana Capac, the Great Inca whose sons, Atahualpa and Huascar, were contesting the crown at the time Pizarro arrived in Peru, the invaders found great piles of golden furnishings and ornaments in addition to 100,000 pounds of gold ingots, each ingot weighing about five pounds.

The swift Spanish conquest of the Inca Empire benefited from internecine strife. The empire had been torn by the struggle between Huascar and his half-brother Atahualpa. Huascar was an iconoclast who wanted to strip the dead of their gold. Shortly before Pizarro arrived, Atahualpa had captured Huascar.

The conquistadors exploited the prediction made by Huayana Capac before his death that white men, sons of the great creator god *Viracocha*, who had long ago disappeared into the sea, would return and conquer the Incas. Pizarro ambushed Atahualpa and his five thousand followers at Cajamarca; two thousand Incas fell, their white cotton tunics stained crimson. The Spaniards kept Atahualpa prisoner, dined with him, allowed him his harem and waited as his promise of gold for the white gods was fulfilled.

Atahualpa, noting the gold lust of the invaders, had conceived of a plan for his release. He promised to fill a chamber twenty-two by twenty-seven feet with gold as high as he could reach. Pizarro agreed and gave him two months. From Cajamarca runners went out to every corner of the kingdom requesting ransom. Immediately porters thronged the roads bearing gold to Cajamarca great hammered plaques, animal figurines, vases, pots, dishes, masks...every conceivable object wrought in the precious metal.

Gold artifacts, the fruit of centuries of craftsmanship, piled up before the wide-eyed Spaniards. They amounted to about 13,000 pounds of gold, It was the greatest ransom ever accumulated. One contemporary chronicler reported that no fewer than 100,000 cargas (litters), borne by four men each reached the town. Pizarro considered Atahualpa's throne and litter the single most precious item in the

73

ransom. It was made of 16-carat gold paved with clusters of emeralds and other gems and required 25 men to carry it.

Large figurines of gold and silver were plentiful in 16th-century Peru where they represented ancestors. These venerated guacas, or totems, were among the objects most pleasing to the Spaniards, not for their artistic qualities, but for their intrinsic value in gold. Each ruler also had his own guardian deities called huauqui. One of these made of solid gold was part of the ransom — a figure so large that it had to be broken up for transport to Cajamarca. Indian goldsmiths were forced to melt down all of the treasures in specially built clay furnaces and cast them into easily divisible ingots. There was so much ransom it took them a full month, ten times as long as the melting down of Montezuma's treasure in Mexico. In thanks the Spaniards strangled Atahualpa and gave him a fine Christian funeral at which the priest who had urged his execution intoned a solemn mass for the dead.

While Atahualpa's ransom was piling up, some of Pizarro's men were already raping the country for whatever treasure they could find, terrorizing and killing. As soon as the Inca was dead the flow of treasure ceased. Runners on the highways, on their way to Cajamarca, got the word and disappeared. What happened to the gold they were carrying? Many believe the treasures still lie hidden in mountain caves and bottomless lakes.

According to one story, there was a Quito businessman named Suárez in the late 16th century who had been kind to one of his servants, an Indian named Cantuña. When Suárez found himself in dire financial straits, Cantuña told him to construct a hidden smelting furnace and then disappeared. He returned a few nights later with 100,000 pesos' worth of gold plate and ornaments; more deliveries followed. The Spaniard died leaving all his wealth to Cantuña. Suspicious authorities questioned the servant who swore he sold his soul to the devil in exchange for an endless supply of gold. He died without revealing the source of the treasure.

Something else the Spaniards never got in the ransom was the famous Chain of Huascar which had been commissioned by Huayana Capac to celebrate the birth of his heir, Huascar. Indian informants told the Spaniards that it took two hundred dancers to carry the seven-hundred-foot "great woolen chain of many colors, garnished with plates of gold, and two red fringes at the end." The chain, life-sized statues, animals in gold, gold bars, plate and ornaments — described by the Indians but never found by the conquistadors — are sought today by modern treasure hunters in the Andean wilderness.

The Inca capital was the richest prize of all; nowhere else in the New World was such astounding wealth found. Pizarro and his men marched on Cuzco, making golden horseshoes for their horses as the iron ones wore out. A contingent of soldiers had gone ahead to the Andean city and reported the most dazzling sight yet seen. The Great Temple of the Sun in Prescott's words was "literally covered with plates of gold. The number of plates they tore from the temple was seven hundred; and though of no great thickness, probably they are compared in size to the lid of a chest 10 or 12 inches wide. A cornice of pure gold encircled the edifice, but so strongly set in stone that it fortunately defied the efforts of the spoilers." The soldiers sent back the embossed gold as part of the ransom, supervising its removal by captive Indians.

Inside the sanctuary the golden-clad mummies of the Great Incas were enthroned with golden masks and crowns. At one end of the lofty chamber was the figure of the sun, a great disk with flaming rays made of a double thickness of gold. Outside the temple was one of the incredible Inca gardens whose like has never been seen in any part of the world where grazed 20 gold llamas. The splendid flock was guarded by shepherds of the same precious metal. Nearby was a gold-plated fountain where sacred corn drink was kept.

Pizarro entered the imperial city a year to the day after he had taken Cajamarca. Even though the temple had already been stripped of its other treasures, the city yielded far more than Atahualpa's ransom. Sent back to Spain, the fantastic amount of gold easily worth five million dollars today had a sharp and immediate effect on prices, sending them soaring. Although insensitive himself to artistic genius, Pizarro sent the King examples of Inca goldwork. Not one of the ten full-sized statues of women, four life-sized llamas, a "cistern" of 38 solid gold vases is known to have survived the melting pot. The crown desperately needed every ounce of bullion and could ill afford to collect art.

The walled city of Cuzco was a blend of the sophisticated and the simple. Low, windowless, shock-resistant buildings, thatched with straw, were laid out on streets at right angles. In the dim light of the interior, gold and silver wires were interwoven in the thatch and gold and silver plates sheathed the walls. The mansions of the rulers were built of stone so carefully fitted together no mortar was used, although sometimes molten lead, silver, or even gold was used to fill the cracks, which explains why the Spaniards tore down so many walls.

Garcilaso de la Vega, son of a pre-Conquest Inca princess and Spanish captain, was the most widely known writer on Peru. His Royal Commentaries of the Incas describes the sacred buildings and royal chambers which were lined with gold:

"In preparing the stone they left niches and empty spaces in which to put all sorts of human or animal figures...all of which were of gold or silver. All the tableware in the house whether for the kitchen or the dining hall was of solid gold, and all of the royal mansions in the empire were abundantly furnished in tableware, so the king need take nothing with him when he traveled or went to war...."

He goes on to describe the fairy-tale gardens whose marvels were melted down with the rest of the new world's treasures:

"In all the royal mansions there were gardens and orchards, given over to the Great Inca's moments of relaxation. Here were planted the finest trees and the most beautiful flowers and sweet- smelling herbs in the kingdom, while quantities of others were reproduced in gold and silver, at every stage of their growth.... They made fields of maize, with their leaves, cobs, canes, roots and flowers all exactly imitated.... They did the same with other plants making the flowers, or any part that became yellow of gold, and the rest of silver. In addition to this there were all kinds of gold and silver animals in these gardens, such as rabbits, mice, lizards, snakes, butterflies, foxes and wildcats, there being no domestic cats. There were birds set in trees, as though they were about to sing, and othebent over the flowers, breathing in their nectar. There were roe deer, lions, and tigers, all the animals of creation."

8 — *Artisans of the New World...*

Treasures

One treasure-laden ship bearing Montezuma's gifts to King Charles V managed to reach Spain. Renaissance Europe was thrilled with the fabulous Aztec treasures which were displayed with great ceremony at Seville, Valladolid and finally at the court at Brussels. The most sophisticated observers were amazed at the forms the New World goldsmiths had coaxed from the yellow metal they called *teocuitlatl* or "the excrement of the gods." Sadly, little remains of these state gifts but the impressions recorded by some of those who saw them.

Peter Martyr, the respected Hapsburg historian and a member of the powerful Council of the Indies, wrote: "But surely if ever the wits and inventions of men deserved honor or commendation in such arts, these seem most worthy to be held in admiration. I do not marvel at gold and precious stones. But I am astonished to see workmanship excel the substance. For I have with wondering eyes beheld a thousand forms and similitures, of which I am not able to write. I never saw anything that might so allure the eye of man."

Among the marvels were gold miters, animals, birds and fishes splendidly crafted to resemble the living creatures, shields of gold and mother-of-pearl, bracelets and neck pieces set with jade and gems, gold bells, a gold scepter studded with pearls and a costume with a mask of topaz alleged to have been worn by Quetzalcoatl, a jaguar skin coat set with jewels, anklets of gold, labrets, pendants and a necklace with 183 emeralds.

Most dazzling was the great "sun all of gold" nearly seven feet in diameter mentioned by Albrecht Dürer. Trained as a goldsmith, Dürer was agog at the Aztec gifts. He noted, "Also did I see the things which had been brought to the king from the new golden land.... And I have seen nothing in all my livelong days which so filled my heart with joy as these things.... I was astounded at the subtle genius of the people in foreign lands."

By the time the Spaniards arrived Mexican goldsmiths were masters at making exquisite miniature articulated gold figurines, working with fine gold wire as opposed to true filigree, cire-perdue (lost-wax) casting and gilding on copper as well as the simpler techniques of repoussé and hammering sheet gold. No one has ever equaled their skill at crafting miniature pieces of such a complex and intricate nature.

The Mexican goldsmiths, above all the southern Mixtecs, were far superior to those of Spain according to a Franciscan who accompanied the conquistadors. He wrote that "they could cast a bird with a movable tongue, head and wings; and cast a monkey or other beasts with movable head, tongue, feet and hands and in the hand put a toy so that it appeared to dance with it; and even more they cast a piece, one half gold and one half silver and cast a fish with all its scales, one scale of silver, one of gold, at which the Spanish goldsmiths would much marvel." The most famous Renaissance goldsmith, the Florentine Benvenuto Cellini, spent weeks vainly trying to duplicate such a flexible polychrome fish after he had seen Aztec treasure at the French court.

The history of goldworking in Mexico began rather late and is puzzling inasmuch as there is little evidence of an archaic or technologically primitive period preceding the flowering of a full range of techniques. Gold was held to be of divine origin. Prized accordingly for its spiritual associations and impervious beauty, it was never employed as a medium of exchange. Indigenous goldwork did not appear until the 10th century A.D., when, rather rapidly, goldsmiths reached a new and higher level of technical skill in casting.

The Mexican smiths, whose work finds its highest expression in the intricate Mixtec cast gold, shunned ostentatious displays of large expanses of shining metal. In this, they were certainly influenced by the fact that there was far less gold in Mexico than Peru, where goldwork at times relied on the dramatic impact of masses of highly burnished metal at the expense of fine workmanship and balance.

Knowledge of pre-Columbian goldwork in Mexico is chiefly derived from the remarkable illustrated encyclopedia of Indian life compiled by a precursor of the modern anthropologist, the Spanish friar Bernardino de Sahagún. His key work is commonly known as the Florentine Codex because the Medici Library in Florence contains the original manuscript. It was compiled in the years after the conquest by Sahagún, who trained the sons of the Aztec nobility to interview informants and transcribe their answers in Nahuatl written with Latin characters. His work details the operations of the goldbeaters, who limited their activities to thinning gold on a stone and polishing it and of the master goldsmiths or finishers. It features many illustrations showing the processes of cire-perdue casting involving clay, charcoal, beeswax and alum.

The Aztecs took captive Mixtec smiths from Oaxaca to work in the capital. But Aztec goldsmiths, unlike those of the Inca who were employed by the state and closely supervised, were generally independent craftsmen. They bartered their goods at a gold market held every fifth day in the capital where there were separate stalls for the purveyors of gold objects and those dealing in dust and nuggets. A small town on the shores of the lake was home to many goldsmiths. Other artisans lived in a special quarter of Tenochtitlán where they were protected by the divinity "Xipe-totec," Our Lord of the Flayed One. At an annual festival those who had been caught stealing gold the preceding year were sacrificed and skinned before this idol.

Systematic grave robbing has been carried out in all areas of the New World where gold was found. It began early. Bernal Díaz, the chronicler who accompanied Cortés, mentions a soldier who extracted some 50 pounds of gold and a great many jewels from the tombs of chiefs in the first recorded instance of American grave robbing. The rapacious Spaniards were so intent on turning the wrought gold of the subjugated peoples into crude, anonymous bullion that virtually nothing that was not buried escaped them. The examples displayed in museums today are almost all from graves robbed in the 20th century.

Until not too long ago the "guaceros", who take their name from the guacas, or gold artifacts they extract from tombs, sold their finds by weight to men who melted them down and sold them for dental gold, to jewelers and those individuals who require a store of gold to feel secure. The governor of the Bank of England reported in the late 1850s that each year the bank received pre-Columbian gold artifacts worth several thousand pounds in bullion to be melted down. Between 1859 and 1861 in the Chiriqu province of Panama gold artifacts worth at least several hundred thousand dollars were stolen from ancient tombs. One grave yielded 701 gold sun disks. Fortunately, today even the simplest guacero realizes the enhanced value of such pieces if they are sold as artifacts.

Because objects made of gold were sometimes concealed in the noses of stone statues in Colombia, many treasure hunters, while not destroying worked gold, have smashed countless idols. In most countries pre-Columbian artifacts are claimed by the government as part of the national patrimony. In practice, many are illegally smuggled to private collections in Latin America and auction houses, galleries, collections and museums in the United States and Europe.

Although examples of indigenous goldwork have been found in many parts of the New World, in three areas the art of the goldsmith reached peerless heights of technical skill and artistic expression. One, of course, is Mexico. The other two are much broader geographically. The earliest metalwork appears in a region centered in Peru and touching Ecuador and Bolivia. The Chimu of the north coast of Peru who worked marvels in gold between A.D. 1200 and 1450 were the greatest goldsmiths of the area. The third is the vast expanse centered in the rich auriferous valleys of Colombia, where the virtuoso Quimbaya smiths worked. It stretches west through Panama and Costa Rica and east to Venezuela's Lake Maracaibo, where superb objects were made by smiths working with stone hammers and crude wind-draft casting furnaces.

In this last region many different cultures developed, each with distinctive characteristics but sharing a wide range of imagery. The predominant motifs are carefully studied natural forms abstracted into symbolic figures of humans and the creatures of air, land and sea. Whether Tairona, Sinu, Darién, Calima, Quimbaya, Chibcha or Coclé, the pre-Columbian goldsmiths, while varying in particulars, were all expert practitioners of a wide range of goldworking techniques. Their cast pieces, the subjects of which were the familiar inhabitants of the natural world around them, are masterpieces of restrained vitality infused with magic. They reflect a pantheistic oneness with nature; every image contains the essence of the thing it represents.

In some areas the goldsmiths practiced several highly developed techniques working in platinum and in tumbaga. Through a process called *mise-en-couleur*, the surface copper in tumbaga pieces of gold and copper was removed by immersing the object in acid baths. Some specialists conjecture that this highly complex process as well as the techniques of granulation and filigree casting may have been introduced from Southeast Asia. There is speculation, backed by mounting evidence, from the noted Austrian archaeologist Robert Heine-Geldern that metallurgy was introduced into pre-Hispanic America through contact with coastal China after the 8th century B.C. He suggests that trans-Pacific voyages from the Dongson area of Indochina influenced goldworking from Costa Rica to northwestern Argentina. Heine-Geldern theorizes that it might have been knowledge of gold deposits in Peru and Ecuador that stimulated repeated Chinese voyages and even fostered permanent settlements. It appears that complex Asiatic metallurgical techniques may have spread from the Pacific Coast to the gold-rich areas of Panama and Colombia. If so, they never reached the Caribbean islands where the level of goldworking remained primitive.

Among the cultures that made use of tumbaga with its low melting point were the Quimbaya of Colombia. Quimbaya cast-gold sculptures, figurines and ceremonial vessels with their elegant, restrained and sensuous contours rank with the world's finest goldsmithing. The artisans of Darién, in present-day Panama, fashioned spectacular anthropomorphic gold pendants in which the human figure has been highly abstracted. These pieces were evidently widely appreciated and traded to other cultures. One such pendant was excavated from the sacred Mayan cenote of Chichén Itzá in northern Yucatán.

The exceptional goldsmiths along the Sinu River, which flows across Colombia's Western Cordillera into the Caribbean, perfected filigree casting. They specialized at producing extremely tricky hollow cast sculptured figurines in the form of animals and birds as well as large, intricate lace-like earrings. The only Colombian culture which the Spanish documented in any detail was that of the Muisca, sometimes referred to as the Chibcha, who inhabited the highland basin around Bogotá. In 1537 the Spaniards conquered the Chibcha and their gold. At the end of the first day's looting so much gold had been amassed that one mounted conquistador couldn't see another over the heaped treasure.

The Muisca civilization was not advanced; their historical memory went back a mere 60 years. They lived in crude huts and their goldwork was relatively primitive. But they had a great deal of precious metal and the Spaniards, ever concerned with quantity not quality, were impressed with the golden displays in the simple villages. The temples and chief's dwellings, made of saplings and straw, were hung, inside out, with thin gold plates and golden wind chimes. The chiefs were borne on gold-covered litters and adorned with shining ornaments.

All of this gold was imported from neighboring areas. The Muiscas traded salt, emeralds and cotton for gold. They made offerings of small cast-gold effigies in the form of humans, real and mythical animals, and miniature "atlatls," or throwing sticks. One of their gods, "Chibchachun," was the patron of both merchants and goldsmiths. Recently the sites of ancient goldsmiths' workshops have been found and excavated. Among the fragmentary evidence they furnished were strips of gold, gold dust, stone molds for casting, crucibles of refractory clay and chisels and tools of tumbaga.

In Ecuador goldsmiths worked with platinum many centuries before Europeans. In a technique called sintering, later forgotten until the 19th century, fine grains of platinum, which has a melting point of 1,773 degrees centigrade compared with 1,0630 degrees centigrade for gold, were mixed with gold dust. The combined metals were alternately hammered and heated with a blowpipe on charcoal until they fused.

9 — *Seeking a Golden Land...*

Eldorado

I n the middle of the last century an amazing discovery was made in a Colombian lake. Fishermen pulled up a gold raft with five figures representing a chief and nobles on it. This marvelous gleaming object fanned the flames of the famous El Dorado (or Eldorado) legend. For centuries, the quest for Eldorado, or the land of the Golden One, lured adventurers of many nations to their fate. As early as 1530 curiosity of the Spaniards' had been piqued by tales heard in the South American lowlands of a Golden Land and its fabulous Golden King. The stories the Indians told had their origins in a religious ritual performed by a Muisca chief on the sacred Lake Guatavita in the area of Bogotá.

The chief, regarded as divine descendant of the sun, made a solemn sacrifice at an annual festival. At dawn of the appointed day he covered his resin-anointed body with gold dust, donned golden regalia, and was rowed on a ceremonial raft into the lake while his people watched from shore. As the sun broke over the horizon, the resplendent chief cast gold and emeralds into the waters beneath which the sun-god lived and then dove in himself. The actual ceremony ceased about 1480 when the tribe was subjugated by a more powerful one, but the legend of Eldorado, the gilded one, persisted.

Who could doubt that a continent that had yielded so much gold contained a golden kingdom? The intrepid Europeans who persisted in the search for Eldorado from Quesada to the Elizabethan adventurer Raleigh explored much of northern South America in the process, just as North America was plundered by men searching for the Seven Cities of Cibola and Golden Quivira. As the hazy location of Prester John's golden kingdom shifted during the Middle Ages from Asia to Africa, so Eldorado, originally sought in the forbidding ten-thousand-foot-high volcanic lake of Guata vita, was moved to the mysterious regions of the Venezuela-Guyana frontier.

Gold-hungry explorers launched searches from Ecuador, Peru, Colombia, Venezuela, Guyana, and Trinidad Island. In Ecuador one of Pizarro's lieutenants, Sebastián Mojano de Belalcázar, who conquered Nicaragua and later betrayed Pizarro, put together a band of 200 gold seekers, including criminals and scoundrels of every stripe. They made an incredible trek from Quito covering 1000 of hellish terrain. Belalcázar reached the great plain in 1538 to find Spaniards already there.

In the spring of 1536 another expedition, headed by the cultivated lawyer Jiménez de Quesada, had left for Eldorado from Santa Marta on Colombia's Caribbean coast. Quesada's ragged force reduced to 167 emaciated men and 59 fever-ridden horses was the first to reach the reported site of Eldorado in January of 1537 and managed to subdue a large number of Chibchas. They got a great deal of the Muisca gold mentioned above, but Spanish accounts tell of great treasures hidden from all of them, including a life-sized idol toppled into the lake by the Indians.

By strange coincidence a third expedition converged on the almost inaccessible plain at about the same time. Such was the attraction of gold that three expeditions starting from widely divergent points, penetrating hostile unexplored country, managed to reach a remote spot despite indescribable suffering along the way.

The third expedition was unusual because it was organized by Germans. In return for a desperately needed loan, Charles V had

Over

At bottom are 16th-century Peruvian bold bars; above is a conquistador's gold toothpick found by Bob Marx with a metal detector at Old Panama City.

Facing

Bob Marx, the author's husband, displays Spanish gold bars he recovered from the 1622 Treasure Fleet off the Dry Tortugas in 1700 feet of water.

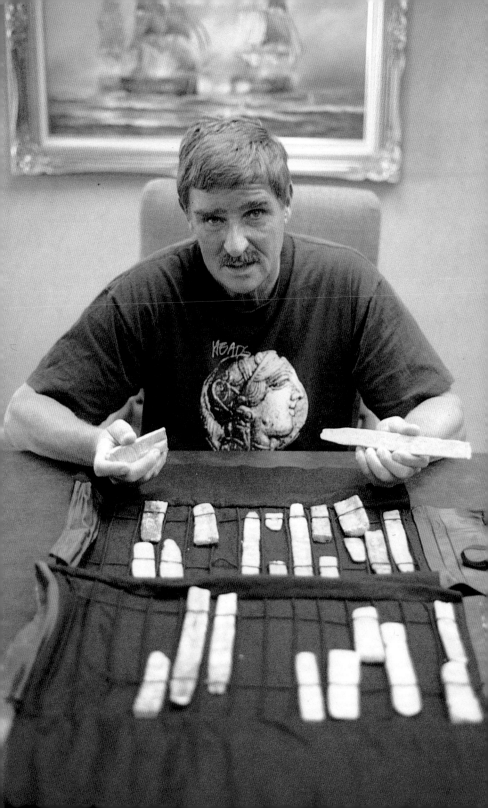

been persuaded to grant the great banking house of the Welsers a patent for conquering and settling Venezuela. The Augsburg merchant princes sent out an explorer named Nikolaus Federmann, one of the few German explorers in the Age of Discovery. For three years he wandered with the remnant of a force of five hundred through the harrowing Orinoco swamps and over ten thousand-foot-high mountain passes in search of Eldorado.

He emerged on the plain of Bogotá with his starving men where he met Quesada, who had arrived six months before and was soon joined by Belalcázar, whose forces like those of the other two had been decimated by poisoned arrows, tropical fevers, skin ulcers, serpents, alligators, jaguars, and hunger.

In 1541 Pizarro's younger brother sought Eldorado. Gonzalo Pizarro crossed the Andes with a party of two-hundred Spaniards, four thousand Indians, five thousand pigs, one thousand llamas, and a pack of bloodhounds. Half the men perished and his supplies were exhausted, but Pizarro, clinging doggedly to the vision of Eldorado, reached the plateau of Bogotá and saw at once that the village of rounded huts was clearly not Eldorado.

The natives indicated the golden place lay beyond the mountains toward the impenetrable wilderness of Venezuela. Not much later, Quesada's brother became the first man to go beyond the mountains. He returned to Bogotá after a two-year quest for Eldorado in Venezuela with only 25 men left of the three hundred Spaniards and fifteen hundred Indian porters who had set out so hopefully with visions of a dazzling city and its fabled King drawing them on.

Over

This magnificent emerald cross crafted in the New World never reached its destination in Spain but was found centuries later on a shipwreck in the Bahamas.

Facing

Pizarro's conquistadors encountered such beautiful golden objects in Peru as this ceremonial knife from the Chimú culture of the 11th and 12th centuries.

The Welsers sent another German agent into Venezuela. After slaughtering large numbers of Indians he was himself slain. Two years later a Welser gold hunter named George Von Speier looking for Eldorado explored almost to the equator in upper Brazil. The last German expedition of the Renaissance ended when a knight was slain in a village in the Orinoco jungles rumored to be filled with gold statues. By the time that the tale of Eldorado had migrated to the Venezuelan jungles it had been greatly embellished. The village on Lake Guatavita had become the fabulous city of Manoa where the streets were paved with precious metal. Over the years, scores of adventurers were seduced by the enchanting tale of the Land of Gold.

Sir Walter Raleigh

One man who had succumbed to the lure of Eldorado was Sir Walter Raleigh, who lost his head, figuratively and then literally. His 1596 book The Discoverie of the Large and Beautiful Empire of Guyana, was widely read and fired English dreams of a golden land to rival Spain's. Raleigh was a conquistador after his time; a man obsessed with gold who overvalued its importance to a nation developing a stable economy based on trade and cloth manufacturing. "Where there is a store of gold it is in effect needless to remember other commodities for trade," he asserted. His dream was of "discovering a better Indies for her Majesty than the King of Spain hath any."

"It is his Indian gold," Raleigh wrote, "that endangereth and disturbeth all the nations of Europe." A courageous but improvident man, he sunk his entire fortune and that of his wife into the quest for Eldorado. Raleigh believed abundant gold would give Protestant England financial superiority over Catholic Spain and quickly wither the Spanish sinews of war. Poor Raleigh, already out of favor with the Queen, returned from his first expedition with pathetic samples of gold ore he had carved from a piece of quartz. His detractors claimed that he had never even crossed the Atlantic but had hidden in Devon or sailed to the Barbary Coast and traded for the gold. The ore brought back was assayed and pronounced worthless. With a stubbornness that matched Columbus's hopes for Hispaniola and Cuba the quixotic adventurer persisted in believing in the limitless wealth of Guyana. Ironically Guyana has been a steady source of gold during the late 20th century.

He continued to pursue Eldorado; the quest had become a fever raging in his brain. He was positive that "the Prince that possesses that land shall be lord of more gold than that gained in Peru, Colombia or Mexico," He explored further, still convinced the land was full of gold mines, rivers, and tombs, but came back empty-handed. In 1616 Raleigh was released from his long imprisonment in the Tower to bring back gold from Guyana on one last try. Again he failed. His son was killed on the disastrous expedition, and his chief lieutenant committed suicide. Two years later Raleigh, a sick and broken man, was beheaded on Tower Hill.

The myth of Eldorado lived on, appearing frequently in European literature. Voltaire's Candide traveled there and described children at play with golden quoits. On 17th century maps of all nations it was vaguely marked and labeled "the largest citie in the entyre world." In 1637 two Franciscans combed the east slope of the Andes for a "Temple of the Golden Sun." Although a number of geographers had openly begun to doubt the existence of Eldorado, the Portuguese mounted an expedition from Brazil exploring north for the fabled land, and in 1714 the Dutch West India Company sent an exploratory force to find Eldorado, or Manoa, a nonexistent city on a nonexistent lake in Venezuela.

At the end of the 18th century the great Prussian scientist Alexander von Humboldt made a thorough study of the Eldorado myth. He followed earlier dream seekers, tracing a route through swampland, jungle, mountains, along the Orinoco until he reached Colombia's Lake Guatavita. The German naturalist reasoned that if Quesada had found gold worth 4,000 pesos de oro when he had the lake partially drained in the 16th century, there must be much more treasure there. Humboldt came up with an estimated figure of 500,000 gold pieces, which was printed in papers around the world inspiring dozens of lake drainage schemes.

In 1910 Contractors, Ltd., of London made an ambitious recovery effort, shipping $150,000 worth of equipment to Bogotá and then draining Lake Guatavita, which was already very low as the result of long drought. Indian gold worth $10,000 was recovered by the group, but they failed to make a profit, since expenses ran over $160,000.

North America had its share of legendary gold. The Seven Golden Cities of Cibola, the kingdom of Quivira, California, a golden land of Amazons, and other fables circulated widely during the Age of

91

Discovery. The image of shining cities with golden buildings four and five stories high kept Hernando de Soto and his men slogging ahead on a relentless four-year march that covered over 350,000 square miles of unexplored North America. De Soto died on the shores of the Mississippi without finding any gold.

At the same time Coronado searched the Southwest for the golden cities of Cibola. He discovered that the towering golden buildings men spoke of were the Indian pueblos of Arizona and New Mexico whose pale adobe walls gleamed golden under the setting sun. Coronado then turned north, pursuing another golden realm called Quivira. He never found it but Spanish chain-mail armor has been excavated from the remains of the Quiviran Indian villages on the Kansas prairies. By 1543 other expeditions looking for the Cities of Gold had explored along the Gulf of California and carried the Spanish influence as far north as Oregon.

Empire in Peril

The disastrous effect the gold of the gods ultimately had on Spain might be considered ample revenge for their 16th-century pillaging and annihilation of entire civilizations. Following the discovery of the New World with its promise of endless bounty, no one would have guessed that so much gold and silver could bring anything but wealth, power and national prosperity. It seemed inconceivable that the American treasure would spark an inflationary spiral that would leave Spain as one of the continent's poorest countries and have harmful effects which still persist today.

For a number of years following the conquest of Mexico and Peru their gold and silver greatly enriched Spain. A current of energy was felt in almost every sector, particularly in literature and the arts which enjoyed a tardy renaissance. The genius of Lope de Vega, Cervantes, El Greco and Velazquez was acclaimed throughout Europe and distracted attention from the adverse effects of the American treasure. In time, however, the aura of prosperity fostered a debilitating decadence. The rise in domestic prices generated by the treasure was successfully coped with at first, but before long a disturbing volume of gold and silver was going to pay other nations for their goods.

Spain found herself in the trying position of maintaining an impossibly large empire, battling both Protestant and Moor and losing most of the gold and silver for which she so dangerously strained her resources. Overall in the centuries of Spanish rule, only about 10 per cent of America's vast gold production actually came to the crown. Such colossal amounts of treasure produced by slave labor appeared to be a great blessing and made Spain the envy of all Europe. Only a very few contemporary economists, among them a group at the University of Salamanca, had the insight to warn of the repercussions of such a massive influx of bullion into an economy that made little parallel effort to develop industry. Throughout the 16th century commodity prices in the country soared, rising with each new shipment of American treasure.

The Spanish crown was perennially in debt to the great banking houses of Germany, the Low Countries and Italy. On a number of occasions banking firms were ruined when the King suspended payments. By the latter part of the 17th century the country had almost no exports to attract foreign capital. Some years fully two-thirds of the New World treasure production was mortgaged in advance to foreign creditors and never entered the country.

The loss of so much bullion combined with the expenses of maintaining the colonies drained the nation of economic vitality and worsened inflation. The royal preference shown to the merchants of Seville and the loss of so many young men to the colonies weakened the country further. Each monarch added to the mounting debt and passed it on to his successor. The currency was repeatedly debased with the result that prices went up even more. The crown defaulted seven times between 1557 and the 1680s when a chain of royal bankruptcies flared into widespread rioting.

New World treasure inspired Counter Reformation ambitions in the Catholic sovereigns of Spain. The Hapsburgs made their country the center of Catholic military operations but failed to stem the forces of Protestantism. Despite every attempt on the part of the crown to maintain their official monopoly on trade in the colonies, the economy of the two began to separate. England and Holland found increasing commercial profit in clandestine dealings with the colonists who were delighted at the lower prices of their goods. Pirates and privateers grew bolder, preying upon Spanish shipping and making incursions into Spanish coastal settlements to strip them of treasure. All of these factors contributed to the ruin of the Spanish Empire.

Although modern-day historians and economists differ greatly in calculating the total amounts of precious metals extracted from the New World, Alonso Morgada, in his history of Seville published in 1587, wrote: "the bullion from the New World which has come into Seville, is enough to pave the streets of Seville with blocks of gold and silver." Peter Martyr writing in 1516 stated that the annual production of gold in Hispaniola exceeded 400,000 ducats. Yet, by 1550 the Indian treasure hoards were exhausted and extensive mining had been organized.

The volume of silver reaching Spain was probably about five to ten times as great as gold and there were in addition precious gems and pearls. Whatever the exact figures were for the gold mined and carried to Spain, the net result was that the mineral wealth pumped into the country altered the political balance of Europe, gradually weakening and impoverishing non-industrial Spain and boosting the economies of manufacturing countries such as England and Holland.

As the chief supplier of European gold, until Brazilian gold was discovered in great abundance by the Portuguese in the late !7th century, the Spanish crown rigidly controlled the production of New World gold to the detriment of her own colonists. Licenses were necessary for emigration to America and concessions sold for mining rights and collecting gold under the *encomienda*   system. All gold, by law, had to be delivered at regular intervals to the royal assayers, where it was refined, scrupulously weighed and stamped...with the King's portion then removed.

Although silver coins were being struck in Mexico City as early as 1536, the first gold coins were not stamped until 1621 in Bogotá. Later gold coins were also minted in Lima, Cuzco, Mexico City and Guatemala. Royal officials were entrusted with the frustrating task of preventing contraband gold from being smuggled aboard the homeward-bound ships. All treasure cargoes were restricted to landing at Seville. Treasure was immediately transferred to the House of Trade, seat of the organization that had total authority over all intercourse between the mother country and the New World colonies. Generally after the treasure was cross-checked with the ships' manifests, most of it sailed off again to pay Spain's numerous creditors.

Seville was also the main commercial center of Spain and the base for the wholesalers who supplied the goods needed by the colonists, who were forbidden to produce all that the mother country could supply such as wine, olive oil, figs, cloth, iron, hardware, paper, books, tools and weapons. The Seville merchants charged outrageous prices since they had no competition, at least not until Europeans defied the Spanish monopoly and sold to the colonies at much lower prices.

The returning galleons brought back other commodities in addition to treasure such as spices, dye woods, indigo, cochineal, sugar, tobacco and hides. They also carried Eastern luxuries transshipped from the Manila galleons which landed at Acapulco. The Asian treasures included gold and silver jewelry, gem stones, ivory, jade and great quantities of silks, spices, porcelains, medicinal preparations and tea.

During the first half of the 16th century small vessels were used to carry the precious cargoes between the New and Old Worlds. By the middle of the century, after freebooters had managed to capture a number of valuable treasure-laden ships, the crown introduced the famous galleons and passed laws stipulating that all returning ships had to sail in convoys, known as "flotas." By the end of the century there were three flotas sailing to and from the New World each year, and the traffic increased from an average of 50 ships in 1550 to more than 150 by 1600.

The flotas made scheduled stops at such ports as Nombre de Dios and Portobello in Panama, Cartagena on the Caribbean coast of Colombia, and at Vera Cruz in Mexico. They regrouped at Havana and sailed together back to Spain. Although many attempts were made to intercept and capture the rich galleons, only the Dutch admiral, Peit Heyn, was successful in capturing a complete flota, a feat he accomplished in 1628 off Matanzas, Cuba.

The loss of ships due to storms and faulty navigational practices was far more crippling to Spanish shipping than pirates or privateers. Over the centuries hundreds of the lumbering treasure galleons wrecked on the treacherous reefs of the Caribbean and many more were sunk in hurricanes. When the ships went down in shallow water, Negro and Indian slave divers were employed to salvage the precious cargoes. Often, however, by the time salvors reached the area of a wreck, it had already broken up and been covered by shifting sands. Many of these wrecks have been discovered in recent years by scuba divers. Appreciable amounts of treasure and many beautiful artifacts have been recovered from them.

Foreigners were also active in trying to salvage the treasures of the Spanish galleons. Wracking, as the profession was called, was one of the main occupations of places like Port Royal, Jamaica and Nassau in the Bahamas. On occasion foreign interlopers were fortunate to find Spanish gold without any effort. In 1724 sailors on a British merchant ship sailing between Florida and the Bahamas spied what appeared to be an unmanned longboat. Upon closer inspection it proved to contain the desiccated bodies of four Spaniards whose ship must have sunk. Aboard the boat the Englishmen discovered a sack of gold dust weighing 83 pounds, a chest containing 4,700 gold doubloons, another filled with silver plate and a small bag of pearls.

In 1678 a Dutch vessel investigated the scattered remains of a ship on a deserted cay near the Caicos Islands. Ashore they found the bleached skeletons of more than two hundred Spaniards who had survived a shipwreck only to perish of hunger and thirst. Evidently a part of their cargo had been salvaged, for the Dutch found more than three tons of gold in specie and bars, 43 tons of silver bars and 340,000 silver pieces of eight on the cay. The lone survivor couldn't explain the tragedy. He was a small dog who was taken back to Holland by the sailors who all became rich men. An expedition was sent out by the Dutch to relocate the wreck and see if it contained any more treasure, but, as often happened, it could not be found.

Everyone rejoiced when the flotas reached home. The arrival of ships that sailed up the Guadalquivir to Seville was scrutinized by ambassadors and economic spies of governments and of the great merchant-banking houses of Europe. On many occasions their arrival saved the Spanish monarchs from imminent bankruptcy and disaster.

In a Fugger newsletter written in 1583 we read: "there came word that the fleet from the Spanish India, praise be to God, arrived without misfortune. It carries a consignment of 15,000,000 [ducats] in gold and silver and people say that they unloaded and left a million in Havana because the ships were too heavily laden. It is a pretty sum and will give new life to commerce."

The Venetian ambassador in Madrid wrote to the Doge nine years later when another flota reached Seville: "After the arrival of the Indies fleet His Majesty has ordered a revision of the account for the West Indies and especially for Mexico, with the result that he finds he hath been robbed of upwards of five million ducats in gold." No doubt many heads were lopped off when the royal investigation was completed.

Pirates, privateers and adventurers of many nations had been attracted to the lure of the New World treasures from the outset. The

97

ship carrying Aztec gold to Charles V had been captured by the Florentine pirate Giovanni da Verrazano, who was sailing under patent from the French king, Francis I. Besides the great amount of gold, the pirates nabbed 608 pounds of pearls and a large amount of emeralds, one reported to be as large as a man's palm. The French King refused to give back the spectacular haul. He scorned the Treaty of Tordesillas which divided the world between the Spanish and Portuguese, declaring "the Sun shines for me as well as for the others.... I should very much like to see the clause in Adam's will that excludes me from a share of the world."

The French, English and Dutch of the Reformation all considered Spanish shipping fair game and they followed in the wake of Verrazano plundering large numbers of Spanish ships over the centuries. Piracy was a national policy employed by many of Spain's enemies in the struggles which had religious, political and commercial motives. Because Spain financed Catholic Europe's Counter Reformation activities it was a matter of vital Protestant policy to sap her strength, and the best way to do so was to keep New World treasure from reaching Spain.

Although the French were first on the scene, their feats were surpassed by the cunning Elizabethan privateers whose Queen endorsed their profitable ventures and excursions in the New World. Elizabeth was badly in need of bullion and faced with mounting inflation. Although she encouraged their depredations, the privateers knew that if they were caught the Queen had no choice but to disavow all association with them.

One of the most successful was Captain Christopher Newport, dubbed "One Hand," having had the other "strooken off" during an attempt to capture a treasure galleon off the coast of Cuba. Newport's third wife was a Glanville, the prominent London goldsmithing family. He had tested his mettle on one of the first English privateering expeditions to the Brazilian coast and also sailed with Sir Francis Drake for a time. After many years spent harassing Spanish shipping and raiding towns in the Caribbean he amassed a considerable fortune and became admiral of the Virginia Company.

The Glanvilles financed many privateering expeditions and bought choice items from the privateers. They got their start in the illicit business through dealing in bullion and gems. They had ships and capital and were set up for evaluating and marketing prize goods. They also had the clout to obtain credit when necessary.

The man who was the greatest Elizabethan seaman and personification of the Elizabethan spirit was Sir Francis Drake. He was a cousin of John Hawkins, one of the first Englishmen to undertake illegal trade with the Spaniards in the New World. Drake had established his reputation as a privateer with a daring raid on a mule train bearing Peruvian treasure across the Isthmus of Panama to the Caribbean port of Nombre de Dios. His small band of men intercepted the treasure caravan and managed to get their booty back to their waiting ships and home to England, much to the delight of his Queen.

When he first landed in Panama he told his men: "I have brought you here to the Treasure House of the World. Blame nobody but yourselves if you go away empty." To the governor of Nombre de Dios he declared that he had come "to reap some of your harvest which you get out of the earth and send into Spain to trouble all the earth."

Drake's next mission took place between 1577 and 1580 and was the first English circumnavigation of the world. Sailing in the Golden Hind and with four smaller vessels, Drake and his men first sacked settlements and took a number of vessels along the coasts of Chile and Peru before capturing the famous treasure galleon Cacafuego (Spitfire). The amount of treasure taken from the Spitfire was so great that Drake had to be content with keeping some 20 to 30 tons of gold and chests of gems and jewelry, jettisoning all the rest. Some two hundred tons of silver were left off the Isla de Plata off the coast of Ecuador where they may still lie.

Although they already had a king's ransom in gold, the English privateers made a stop along the northern coast of California where his men scratched about finding "a reasonable quantitie of gold and silver" almost three hundred years before gold was discovered at Sutter's Mill.

The ostensible object of the circumnavigation had been to establish trading relations with the leaders of the treasure and spice states of the Far East. Drake did load six tons of precious cloves in the Moluccas, but almost all of it was lost when one of his ships ran aground on a reef. In spite of this, he returned to England with such colossal amounts of Spanish gold and treasures that he became a very rich man and returned a dividend of 4,700 per cent to the shareholders who had financed the expedition.

The Queen rewarded him with knighthood and he went on to attack other Spanish possessions in the Caribbean — first, sacking Santo Domingo where the Venetian ambassador in Madrid wrote to the Doge that he had captured "1,500,000 in gold booty," then capturing and sacking Cartagena and St. Augustine in Florida.

In April of 1586 the ambassador wrote another letter to the Doge stating, "News from Lisbon and Seville brings identical news on Drake, that he had landed troops and seized Santo Domingo, Puerto Rico and Havana; this latter a real disaster as the Duke of Medina Sidonia told me that they had two millions in gold there." Four months later he wrote again saying, "We just learned that Drake has returned to England with thirty-eight ships laden with much booty." This correspondence gives an indication of how closely Drake's exploits were followed. This great haul paid for the defenses which were later to repeal the Invincible Armada in 1588, and much of England's court plate and jewelry was made from this stolen gold.

In an effort to prevent Peruvian treasure from reaching Spain, Drake was sent in 1595 by Queen Elizabeth to capture Nombre de Dios and Panama City and hold them for ransom. Although Nombre de Dios was captured, along with a considerable amount of treasure, Drake fell victim to a fatal bout of dysentery and was buried at sea in a lead coffin off Portobello. The man whose name struck fear in every Spanish American heart had opened the door of the New World to future colonization by English and other European nations.

11 — *Portugal's Search for Gold...*

Brazil

The Italian navigator Amerigo Vespucci, while searching for an eastern passage to the East Indies in 1501, had sailed along the Brazilian littoral, stopping at each cape and harbor, claiming them for the Portuguese monarch. He encountered various groups of Indians who seemed to prize feathers above all and when offered a gift of a gold cross usually preferred to take a wooden comb or a mirror. The natives told him "wondrous things about the gold and other metal," but the overly cautious Vespucci remarked, "I am of those like St. Thomas who are slow to believe." Neither he nor the government was optimistic about the area's potential, and almost two centuries went by before the Portuguese began to tap the great gold wealth of Brazil.

Portugal claimed Brazil under the terms of the Treaty of Tordesillas but considered the vast land of little value and made scant effort to colonize it. In the early 1500s a small group of converted Jews fled the Inquisition to Brazil, where they exported brazilwood, which was highly valued in Europe as a dyewood. By the beginning of the 17th century there were large settlements along the coast exporting sugar, tobacco, salt and other products. Little by little the interior in the south was penetrated by the Paulistas and the vaqueros, or cowboys. The Paulistas, inhabitants of the highland plateau of São Paulo, were poor farmers of mixed Amerindian blood. Unable to purchase Negro slaves as wealthier men did, they took to the bush of Minas Gerais to capture Indians. During one of these sorties in the 1690s they discovered large deposits of gold and the rush was on.

Throughout the 16th and 17th centuries the Brazilian settlers had found scattered deposits of gold, and some gold had come into Portuguese hands through clandestine trade with Peru. Men who found gold in the Spanish colony sometimes smuggled it into Brazil, where they could sell it without having to pay the onerous Spanish taxes. But such gold activity was desultory and gave no warning of the real rush precipitated by the discovery of the Minas Gerais (General Mines). Discoveries were made almost daily. It seemed as if every rivulet and stream sparkled with gold. For many years all the gold from Brazil was in the form of dust and nuggets washed down from the placer deposits where it had accumulated undisturbed for hundreds of thousands of years.

The Minas Gerais rush, foreshadowing those of the 19th century, drew young and old, rich nobles and illiterate mendicants. They set out with no more than a pack on their back for the new Eldorado. Vast numbers of adventurous gold hunters converged on the large auriferous area on three trails: from Bahia, Rio de Janeiro and São Paulo. Many died of starvation, disease, or in frequent skirmishes with one another. The Governor General labeled them "Vagabond and disorderly people, for the most part base and immoral."

As discoverers, the Paulistas felt cheated by the arrival of such large numbers of men. The interlopers came not only from Brazil with their Negro slaves but from Portugal as well. In 1709 the "War of the Emboabas," as the newcomers were called, flared up, and after several years of escalating hostilities the pioneer Paulistas were chased away. The Portuguese King, Dom João V, sent a governor to rule Minas Gerais. He was to establish a mint there and collect the royal tenth on all gold found,. Colonial officials, aware of the damage that American gold had done to Spain's economy, were both exhilarated and apprehensive. They feared that Brazilian gold might flow into Lisbon and right back out again to pay for imports from other European nations. Their fears were well founded, for between half and three-fourths of the gold that reached the mother country was indeed funneled to northern Europe.

The Paulistas, who had been evicted by the Emboabas, pushed westward through hostile country where they were rewarded by the discovery of the gold fields of Cuiabá, Goiás and Mato Grosso. The word of their finds spread like the wind, attracting hordes of gold-crazed prospectors. According to one source the "good gold," that of twenty-three carats, "made clefts in the nuggets as if it were bursting out on all sides, while from within it gave off reflections that looked like rays of sunlight."

In 1719 a royal decree ordered the establishment of processing plants in all areas where gold was found and forbid the circulation of gold dust as currency. At the mints where the gold was converted into bars and coins, one fifth was removed. Part went to the crown and part to cover "pin money for the queen" and processing costs. When diamonds were discovered the following year in the Minas Gerais district, the entire area was made into a sealed-off crown reserve which could only be entered by presenting an official permit.

By the middle of the 18th century the output of gold from the mines and rivers of Brazil matched that of Spanish America. The chief beneficiary of the Brazilian gold as in the case of the Spanish gold was not the country that controlled it but the most economically developed nations of Europe. England attracted most of the Portuguese colony's treasure with her manufactured goods. A country with virtually no domestic production, she became one of the foremost gold markets of Europe. So many Spanish and Portuguese coins circulated in England that merchants and bankers carried scales designed to equate the foreign pieces with English currency.

The Brazilian gold was carried back to Lisbon in fleets of warships, and their arrival was carefully observed throughout Europe. In London in December 1736 a report was published declaring that "the Brazil fleet, consisting of 14 ships, arrived this month at Lisbon with a rich cargo of 2130 octaves of diamonds and six and a half million in gold cruzadoes, of which four and a half belong to the king and the other two to private merchants." And in September of 1748, twenty-one ships brought back "21,740 pounds of gold, 439,980 cruzadoes in silver and many chests of gem stones."

The discovery of gold and diamonds caused great turmoil in Brazil. Men suddenly abandoned their work on sugar and tobacco plantations or in the coastal towns and flocked to the mines. In addition, prices rose sharply because of labor shortages, and there was a great increase in the slave trade with Africa. The governor of Rio de Janeiro reported to the crown in 1726 that "there is not a white miner who can live without at least one Negress from Dahomey for they say that only with them do they have any luck."

Brazil remained the focus of treasure hunters and pseudoscientific investigations long after the gold boom days were over. The mysterious and tangled heart of the vast land was made to order for persons with visions of finding lost mines and ancient cities blazing with gold. A French gold prospector, Apollinaire Frot, spent many years in the 1920s and 1930s exploring the Brazilian wilds. When he emerged from time to time, he alluded to amazing discoveries, so startling he feared to publish them. Frot claimed to have found a number of ancient petroglyphs in the province of Amazonas which proved that ancestors of the Egyptians had once mined gold in Brazil. One set of hieroglyphs carved on stone supposedly led him to one such mine.

In 1932 a German-sponsored expedition claimed to have found a large city of stone buildings plated with gold and a monumental pyramid filled with gold. They reported they found not only gold but a number of white-bearded dwarfs, but were unable to produce any convincing evidence. Even today treasure hunters still plunge recklessly into the heart of the continent to seek the elusive gold.

Gold from the Pacific

Some of the gold that reached Spain from the New World came via her Pacific territory, the Philippine Islands through the Manila galleon trade. Gold in the form of bullion and jewelry from the Orient was carried east on the Manila galleons to Acapulco, despite Spain's ban on importation of gold manufactures. A contraband gold cargo intercepted in 1767 at Acapulco included Chinese and Philippine objects which included votive images, a large bird, alligator teeth capped with gold, dishes and platters of solid gold, hundreds of gem-set rings and all kinds of jewelry, much of which was studded with diamonds and emeralds.

Over

Four Dutch coins at bottom, dated 1729, were minted from Brazilian gold; above are coins dated 1646, the first ever minted in Brazil.

Facing

The solid gold altar of the Panama City cathedral, was originally built in the 16th century with gold that came from Peru.

In the early years of the trans-Pacific sailings to Mexico, most of the gold came in the form of bartered gold tribute rendered by the natives of the Philippines. Antonio Pigafetta, the chronicler who accompanied Magellan on his voyage of discovery to the archipelago, wrote of alluvial deposits worked by natives in the streams of Visayas. The auriferous sands yielded "pieces of gold, of the size of walnuts and eggs." Miguel López de Legaspi, the conqueror of the islands, wrote to Philip II that the people wore many gold ornaments and that gold was to be found in varying quantities in almost all of the islands. The natives, he feared, were too little interested in hard work to make working the gold deposits profitable to the crown. The Spaniards organized very little mining. They contented themselves with the considerable amounts of gold they could wring from the natives.

The initial payment in gold by two provinces of Luzon alone amounted to more than 110,000 ounces. Thereafter the amounts which went into the Spanish treasury declined. The Spanish governor in 1583 reported that 60,000 to 70,000 pesos in gold were shipped to Mexico in some years. But production was much higher. An increasing amount of gold was siphoned off into illicit trade between the natives and other Europeans who trade with them, such as the Dutch and English.

Over

More Colombian gold that was crafted by an 11th-century culture and dredged in modern times from Lake Guatavita.

Facing

Gold funeral masks; at top from the 15th-cenatury Ecuadorian Tolita culture; at bottom from the Peruvian Vicus culture of 200 B.C.

109

Francisco Martinez de la Costa reckoned that as late as 1783 2 million pesos' worth of gold was mined in the country and traded to foreigners. His figure, however, may be regarded with some suspicion since he also makes the exaggerated claim that Thomas Cavendish, the English privateer, got a prize of 658,000 pounds of Philippine gold when he captured the Manila Galleon Santa Ana. In any case, the Spanish never recognized the great wealth of gold that lay in the mountains and rivers of her Pacific colony. In the 18th and 19th centuries Chinese entrepreneurs organized some gold mining. In the 20th century the United States began large-scale exploitation of rich deposits, particularly on the island of Luzon, which yielded as much as $23 million a year in the 1930s (based on gold at $35 an ounce) and made gold one of the country's leading exports for many years.

12 — In the Age of Exploration...

Gold in Europe

As the 17th century progressed, it became increasingly clear that the beneficiaries of the great golden wealth of the New World mines were neither the Iberians who claimed it, the indigenous Americans whose cultures and very existence were sacrificed to it, nor the masses of West Africans who were enslaved for it. The richest harvest was gathered instead by the manufacturing countries of Europe which had little gold of their own.

The tons of bullion that sailed up the Guadalquivir to Seville passed into the pockets of the northern entrepreneurs, stimulating their developing industries and enriching their mode of living. As they prospered, they joined the nobles and ecclesiastics in patronizing the goldsmith. In the 16th century artisans in such commercial centers as Nuremberg and Augsburg lavished infinite care and detail on an expanding variety of gold and silver-gilt objects.

Germany was the northern leader in goldwork until this area of Europe was ravaged by the Thirty Years' War in the mid-17th century. Distinctive national characteristics developed in the plate and jewelry of Germany, Holland, Flanders and England where merchant princes, princelings and the powerful guild organizations flourished. Goldsmiths added novelties from the New World to their stock of images. Pineapples, coconuts, potatoes, vines, shells and "noble savages" appeared on jewelry and the elaborate sculptured cups then in vogue.

In the late 16th century the Age of Discovery inspired the gold globe cups that incorporated a modeled figure of Atlas, muscles bulging as he supported a terrestrial or celestial globe, or sometimes one above the other. In the hands of a master craftsman, with the Northern commitment to perfection, these were magnificent pieces of sculpture.

111

In the 17th century the richness and florid splendor of the 16th-century Renaissance jewelry with its colorful enamels and baroque pearls gave way to the age of the diamond. Italy no longer set the fashion. France became the arbiter of elegance, grace and refinement.

Lapidaries, particularly those of the Low Countries, made great advances in the faceting of gems. The rose-cut diamond replaced the earlier table cut, revealing the stone's true brilliance for the first time. The diamond became the queen of gems and the emphasis in jewelry shifted from the soft glow of precious metal to the brilliant dazzle of the stones, although enameling remained very popular.

The Italian Renaissance motifs of elaborate swags of fruit, masks and mythological creatures were replaced by French-inspired floral and vegetable themes, delicate bowls and scrolls. The French used gold to great effect on wallpapers, architectural elements and particularly the ornate gilded furniture of the period.

Less jewelry from the war-torn 17th century survives than from centuries preceding or following it. This is in part because Renaissance jewelry relied heavily on enameling and had few stones. It could be broken into smaller components without destroying its beauty. However, with the new vogue there was less reliance on the intrinsic beauty of gold and more on gems, which could easily be removed from their settings and remounted in a more fashionable mounting. Faceted stones could even be recut.

Until the 17th century a goldsmith had been both jeweler and worker in precious metals, using his own designs and cutting his own cabochon stones. With the new emphasis on faceted stones, the crafts separated. The goldsmith chiefly worked on ceremonial and domestic plate. The jeweler, using his own designs or copying from others which were circulated internationally in book form, created jewelry and accessories of gold set with gems, which he purchased from an independent lapidary who cut and polished them. Even enameling became a distinct craft.

In 17th-century England the crown ordered a series of forced sales of plate and jewelry to finance military campaigns. This accounted for the destruction of silver and gold works from that period and earlier. In 1627 King Charles I levied a loan of 120,000 pounds sterling on the City of London to finance the war effort. This was a huge sum and the Worshipful Company of Goldsmiths, which had a large collection of precious jewelry and goldwork, was forced to relinquish many pieces then and more later. By the beginning of the 18th century they had little left. In 1641 the King ordered all plate melted down at the Royal Mint. Not only the prosperous guilds but the nobility, the goldsmith's traditional patrons, were forced to hand over their personal possessions for the King and Parliament party.

In the 18th century European goldsmiths following trends set in Paris gave free rein to their imagination in the execution of small personal accessories such as snuff boxes, watches, fobs and the chatelaines they hung from. The baroque heaviness of the Renaissance was replaced with jewelry of airy lightness elegant and delicate.

Gold as a commodity came into its own in the 18th century and more of it was coined in Europe. Up until this time there had not been enough gold in Britain to institute a gold standard. But as Spain and Portugal pressed slaves to extract precious metals in distant lands, vainly counting on the American treasure to resolve increasingly intricate economic and political problems, the English had embarked on a road that was to lead to the Industrial Revolution. The pound sterling had served as the basis of the English monetary system since Saxon times, but increased trade with the Orient was draining much silver to the East where its value was appreciably higher. At the same time there was a tremendous increase in the amount of gold in Britain. It came not only from Brazil but after 1750 also from Russia.

In 1717 Sir Isaac Newton was Master of the Mint. As a mathematician he made a careful study of trade factors, gold production and international price levels. He slightly lowered the mint price for gold, establishing a fixed price of 84 shillings and 1 1/2 pence per troy ounce. In 1816 the country formally went on the gold standard which had been in effect for almost a century and amazingly Newton's price held stable for more than two centuries until 1931.

In the early 18th century the China trade provided England with a small but interesting source of gold. The merchant ships which returned from the Orient with precious cargoes of tea, silks, porcelains and drugs such as ginseng root and rhubarb often brought gold which had been profitably exchanged for silver. Silver was in great demand in China where it formed the basic medium of exchange. Gold was widely used in the Chinese decorative arts in laquerwork, gilding of furniture and ornaments and the gold threads in sumptuous brocades, but it played no monetary role. In Europe the value of gold to silver was ten to one and even higher in England. English traders made a profit of more than 60 per cent on the deal. Ship's officers shared in the trade. They were generally permitted to leave England with a limited amount of silver to exchange for Chinese gold.

Gold in Russia

Until 1848 it was Russia that led 19th century gold production. The rich pockets of alluvial gold in rivers and steams on the ancient trade routes to the Black Sea and the Mediterranean had been worked since prehistory. They furnished much of the gold that shone so splendidly in the Persian, Greek and Byzantine eras. In modern times the great era of mining was set off by the discovery in 1744 of a quartz outcrop near Ekaterinburg on the eastern slopes of the Urals. Under Catherine the Great serfs panned 84,000 ounces of high-purity gold from the rich alluvial beds in the first four decades of systematic production. The czars sent out prospecting expeditions in the early 19th century, and in the remote Altai mountain valleys west of Lake Baikal they found many new deposits of auriferous sands. This forbidding region of Siberia, haunt of the fabled gold-guarding griffins of Herodotus, yielded vast amounts of gold between 1814 and 1850. Russia was then furnishing the world with over 60 per cent of her newly mined production.

Conditions in the gold fields were unspeakably harsh. The forced labor of the peasant serf recalled the past sufferings of slave miners from Egypt to Mexico and foreshadowed Stalin's 20th century slave gold miners. Working for a small number of feudal lords who had crown licenses or for the crown directly, the serfs toiled on the frozen tundra in a curtain of fog or snow six days a week from 5 a.m. to 8 p.m. They had no rights and existed joylessly on meager rations. European travelers reported that the landlords lived like opulent princes, priding themselves on being able to offer guests the oranges of Sicily, the wines of France and Havana cigars. As a diversion, they escorted visitors to their diggings and consumed champagne and caviar while casually observing the serfs at their labors.

The surge in Russian gold production was reflected in the splendor of the Imperial Court, which set the tone for the lavish displays affected by the Russian nobility. The Siberian gold made possible a proliferation of precious regalia which reflected a passion for diamonds. Empress Catherine wore a crown of gold in which 2,075 diamonds of 1,400 karats completely hid the gold. The best known of all Russian jewelers was Peter Carl Faberge, a virtuoso of French Huguenot descent, whose workshop turned out unique objets d'art for the czars and their families at the end of the 19th-century. Faberge designed fantasies of consummate workmanship, combining gold, enamel (he could produce 144 different shades) and a variety of precious and semi-precious stones to create exquisite objects with subtle rainbow nuances of color.

Part Three

Everyman's Gold

The 19th century was the true age of gold. It began with the continuous working of the Russian Czar's Ural mines and ended with the discovery of the world's greatest gold deposits in South Africa. The Russian gold was mined by serfs who were in effect slaves of feudal landlords and the emperor just as they had been since the Middle Ages. Extracting South African gold, buried deep in ancient quartz reefs, requires massive organization of capital by giant corporations and depends on modern technology.

The frenetic half-century before gold mining on an industrial scale was introduced in South Africa was the era of the free-lance miner. Prizes went to the strong...the persistent...the lucky. An independent spirit and perennial optimism were the hallmarks of those whose courage and determination fueled the most stupendous gold rush in history.

These were the halcyon years of the individual prospector who worked for himself and kept what he found. In California, Australia, New Zealand, Nevada, Colorado, Idaho, South Dakota and finally the incredible Klondike the golden dreams of some men were fulfilled and the lives of countless others broken. Waves of eager men and women swept restlessly back and forth across the world seeking gold. From California to Australia to Alaska they sought the end of their rainbow where

117

fortune beckoned. The saga of these prospectors is filled with stories of sudden riches and heartbreaking failures. Yet fortitude and undiminished hope kept them struggling and searching.

Most prospectors were to learn that every ounce of gold wrested from the earth cost a high price paid in sweat, suffering and loneliness. And, as the 20th century ushered in the era of modern mining few had gained the wealth of their dreams. Little remains of their efforts but ghost towns moldering in the sun and the spirited, often poignant tales from the memory of survivors and their descendants who proudly illustrate them with crumbling journals and letters, yellowing photographs and faded newspaper clippings.

13 — *In Carolina and Georgia...*

U.S. Gold

In the 19th century the wilderness of the New World once again beckoned to the gold seeker. Myth and reality, mingled since the dawn of history, had propelled the conquistadors and their successors through vast uncharted territories on a quest for treasure. Gradually the hazy outlines and features of the new lands came into focus. The Spanish had found a storehouse of gold beyond all expectations, but the ancient dream of the golden land, just beyond the next mountain or deep in the trackless jungle, remained strong.

The English settlers at Jamestown in 1607 were just as much interested in gold as anyone else. One man complained that "There was no talk, no hope, no work but to dig gold, wash gold, refine gold, load gold." In 1608 Captain John Smith sent samples of what appeared to be rich ore to London for assay. It turned out to be worthless. Other prospectors fared no better and gradually high hopes of finding gold in the British colonies faded. In 1790 Benjamin Franklin attested that "gold and silver are not the produce of North America, which has no mines."

Nine years later a boy playing with a bow and arrow near the Rocky River of central North Carolina discovered a 2-1/2-pound lump of gold where an arrow landed. He took the lump home and his father, a man named Reed, made good use of it as a door stop. After a couple of years a friend suggested he might be able to trade it for money. He did and then started mining in the river, collecting some 115 pounds of gold in the next 10 years. Other prospectors made strikes in the same area and large nuggets were found, including one weighing 28 pounds, "of the shape and size of a domestic smoothing iron." In 1821 a mass of almost 50 pounds of gold was found in a rock crevice. A letter to Thomas Jefferson from the director of the U. S. Mint in Philadelphia in 1805 noted that "very considerable" quantities of gold were being produced in the North Carolina gold fields. Gold seekers began to flock to the area in a minuscule gold rush. They came from as far away as Europe, where the find was front-page news.

Until 1828 all of the $138,000 of domestic gold coined by the U. S. Mint came from North Carolina. Quite a bit of the local ore remained in the state, where it circulated in the form of dust and nuggets until a shrewd local businessman decided to open a private mint. Christopher Bechtler, a metalsmith from Germany, had come with his family to Rutherford, North Carolina, in 1830 where he opened a shop to make jewelry from Appalachian gold. His family mint operated until 1852 when many miners had already gone West to try their luck on the richer diggings. But Bechtler coins continued to circulate for a number of years. Because they were a Southern product they enjoyed greater confidence than "Yankee" gold. It was not unusual for contracts from the Confederate States of America to specify payment in Bechtler gold rather than Confederate paper.

Small amounts of gold were washed from the streams of Virginia, South Carolina, Tennessee and Alabama, but the second mini-strike of the century was in northeastern Georgia where gold was found on Ward's Creek. Local lore has it that the first big nugget was kicked up by a deer in 1829. This was within the Cherokee Nation — the area which had been guaranteed by the United States in 1785 to the Cherokee people for "as long as the grass shall grow and the river shall run." Within a year of the strike it was reported that "4,000 persons are engaged in gathering gold at the Yahoola mines in Cherokee country and their daily products are worth $10,000." With what was to become a characteristic disregard for honoring treaties with the native Americans, President Andrew Jackson promptly rescinded the federal government's agreement and the hapless Cherokees were driven along the "Trail of Tears" to settle in dusty Oklahoma. The miners moved in to work the Appalachian gold fields, the "golden land of the Yupaha" which had eluded the Spaniards three centuries earlier.

It was during 1539 and 1540 as Hernando de Soto explored the Southeast that one of his lieutenants was told by Indians in northern Florida of a province called Cale where gold was abundant and of another land where men wore gold helmets. De Soto led his men to Cale and there learned of a much richer land called Apalache, seven days farther west. They went there and found no gold. But a captive boy from a distant land called Yupaha which lay "toward the sunrise," told them of his golden land. A woman ruled and the town where she lived was of wonderful size. She collected tribute from many of the neighboring chiefs, some of whom gave her clothing and others gold in abundance. The Yupaha boy told how it was taken from the mines, melted and refined, "just as if he had seen it done," wrote de Soto's chronicler. But the expedition never found Yupaha.

The U. S. Government established mints at Charlotte, North Carolina, and Dahlonega, Georgia, in 1838. Dahlonega, named for an Indian word meaning "yellow metal," attracted throngs of get-rich-quick characters. Most of them remained poor and anonymous. A few Southern families, however, made fortunes from gold mines which had produced an estimated 50,000 pounds by 1850. One of the best known was that of statesman and orator John C. Calhoun, whose family mine near Dahlonega yielded over $4 million in pre-Civil War gold. Calhoun had a vital personal interest in the perpetuation of slavery, for his mines were worked by slaves. When the Civil War ended, the Calhoun Mine, already close to exhaustion, was forced to close. In total, Appalachian gold deposits yielded less than $20 million in bullion, no more than $300,000 annually. The majority of the alluvial deposits were worked out by 1847 and by 1866 the remaining quartz mines had also been exhausted.

California

The little boom in the southeastern United States was a prelude to the great gold rushes of the second half of the century. The dramatic discovery of gold at Sutter's Mill on January 24, 1848, sparked the California gold rush, one of the greatest mass movements of population in history. The discovery of gold in the wild, largely unexplored and unpopulated territory came less than 10 days before the United States acquired California from Mexico by the peace treaty of Guadalupe Hidalgo on February 2, 1848.

It was not the first find of precious metal in California. Scattered strikes had been made by Indians, Spaniards and others in the centuries before Mexico gained independence from Spain in 1820. Cortés had explored Baja California for the gold that was linked in the Spanish imagination with the island of "California." One of the manifold legends of mystic golden lands, this was the concoction of medieval romancers, an island "lying on the right hand of the Indies," peopled by magnificent Black Amazons who rode domesticated wild beasts adorned with gold. Montalvo, the Spanish fantasist, wrote of their bare-breasted Queen Calafía in whose land there was no metal save gold.

Cortés found no such "California," and one by one, through centuries of fruitless search, the shimmering myths so deeply rooted in the Spanish mentality lost their luster: the will-o'-the-wisps of California, Eldorado, Quivira, Yupaha and the Seven Shining cities of Cibola. No treasures to match those of Montezuma and the Incas were to be found again in the New World.

Of all the dream-seekers' goals only one would fulfill its golden promise. California turned out to be no island realm of hostile women, but a land exceedingly rich in precious metals. In the last 50 years of the 19th century California and the West produced more

123

gold and silver than any similar area before in the history of mankind.

In 1769 Spain occupied Alta California, and the padres founded the first of the California missions. Bartering Indians brought alluvial gold in willow baskets to the priests who apparently were more interested in saving souls than getting rich. They accepted the gold but made little effort to find its source. Europe was dimly aware that there was gold in California; an 1816 English book on metallurgy noted in detail the gold found in the mountains there. The pioneer trapper Jedediah Smith panned alluvial gold in the Sierras a decade later and in 1840 Richard Henry Dana wrote of California gold in Two Years Before the Mast. In 1842 a third-generation Californian, Don Francisco Lopez, dug up some wild onions for his wife while he was out hunting in the hills above Los Angeles. Tiny flakes which he recognized as gold clung to the roots. For a number of years his family quietly and successfully collected gold in the hills. In 1843 a prosperous Los Angeles merchant shipped a packet of gold to the U. S. Mint in Philadelphia where it was coined into the first California gold coins. But it aroused little interest because California belonged to Mexico.

James Marshall's celebrated discovery of gold at Johann Sutter's millrace on the American River came at a fortuitous time for the adolescent United States which was making the painful transition

Over

Gold coins and nuggets, at bottom, were found with metal detectors in the San Francisco area; jewelry and objects, above, were also found with detectors.

Facing

This modern-day prospector follows in the footsteps of the 49ers as he pans for gold along the Yuba River in California.

124

from an agrarian to an industrial age. The country in the 1840s was in a tumult. Politics were corrupt. The Panic of 1837 had left banking and investment precarious. The issue of slavery impacted all aspects of life. Mass immigration, averaging around 100,000 a year, reached 296,000 in 1848, the year of the Irish potato famine, and dangerously strained the national economy which was already reeling after the Mexican War. Gold was in short supply. News of the gold find in California threw the nation into a frenzy. Dreams of American gold were reborn. One journalist commented, "The farmers have thrown aside their plows, the lawyers their briefs, the doctors their pills, the priests their prayer books, and all are now digging gold." He was not exaggerating. Tens of thousands of Americans and immigrants converged on the Pacific wilderness to seek their fortunes. In the process they gave a tremendous boost to the country's economy and morale, supplying an average of 175,000 pounds of gold annually by 1851.

That fateful first gold nugget from the American River gave no clue to the magnitude of what was to come, for it was no bigger than a pea. James Marshall, who started it all, was an itinerant jack-of-all-trades, hired to construct a sawmill for the landowner Johann Sutter. Marshall was a moody and ineffectual man, born under an unlucky star. "My finding gold," he said, "was to deprive me of my rights of a settler and an American citizen." His homestead was overrun by

Over

Metal detector hobbyists also track the path of old-time prospectors as they search for nuggets and deposits of placer gold that can be panned.

Facing

Jewelry made of gold has been popular since ancient times and is still worn today by men and women everywhere such as this Cuna Indian off Panama.

swarms of prospectors. He abandoned it and took up a pan and shovel himself, but through years of prospecting gold eluded him. His last years were spent in drink and abject poverty. He repeatedly petitioned the California legislature for a pension. Finally it was granted but then cruelly withdrawn when he showed up drunk at a legislative session. Only after he died did the state erect a statue in his honor.

Johann Augustus Sutter suffered an even more precipitous decline and fall. A German-born Swiss who had come to the United States via Hawaii, with a record of bankruptcies and forgeries behind him, Sutter's land grant from the Mexican Government included 230 square miles of lush land in the central valley of California. He ruled benevolently over an agrarian empire of some 142,000 acres, issuing his own coins and trading produce as far away as the Pacific Islands and South America. He ought to have been a Croesus from his trading alone not to mention the gold found on his land. But he was a generous, genial fellow, not much of a businessman and perennially in debt. And, far from enriching him, the discovery of gold destroyed his empire and his life.

When Marshall came to him with the yellow flakes and grains from the millrace, Sutter begged him to keep the find secret at least for six weeks. But, the Spanish proverb explains that "Gold and love affairs are hard to hide." The word was soon out. At first men were skeptical. One of San Francisco's two fledgling papers, the California Star, reported that it was all a "sham; a superb takein as was ever got up to guzzle the gullible."

The news inevitably spread. First, Sutter's mill hands quit to devote themselves to working with a rocker, or cradle as it was sometimes called. They were soon pulling $50 in gold from the river each day and Sutter's other employees abandoned work to join them. More men made their way to the Sutter's land but he was unable to hire any of them to harvest his crops which rotted in the fields, to tend his livestock, milk his cows or finish the mill that Marshall had started.

Like creatures possessed, men took baskets, even kitchen sieves, to gather the gold. A fortunate few made as much as $800 a day in some places, and by the end of the year an estimated $6 million worth had been found. Ruffians camped out anywhere, helping themselves to the stock from Sutter's store, his crops and animals.

Sutter, long a drinker, immersed himself ever deeper in an alcoholic daze. Eventually he prosecuted 17,221 squatters and demanded $25 million in compensation from the state of California which had become part of the Union in 1850.

On May 15, 1855, the highest judge in the state ruled in his favor declaring that the huge territory was indeed his. Jubilation was brief, followed by havoc. A mob of more than 10,000 people, angered by the decision, burned the court, seized the state archives, tried to lynch the judge and plundered Sutter's property and buildings, leaving them in flames. His eldest son shot himself, the second son was murdered and the third drowned on his way to Europe. Sutter, a lonely and broken man, never recovered.

For the following 25 years a deranged, stumbling old man in general's uniform haunted the Congress pleading his case with one legislator after another. The story goes that one afternoon as he sat on the Capitol steps, a group of mischievous boys hailed him with cries of "You've won, you've won, Congress has settled it" Hearing this, the old man rose to his full height for the first time in many years and then keeled over dead, his suffering ended.

There were many treasure hunters deluded at the end of the rainbow. But never in history had so many men actually *found* the tantalizing pot of gold and been free to keep or squander it. In 1845 there were but some 700 United States citizens in California. In 1848 California's population stood at 14,000 (excluding Indians, who weren't counted in those days), of whom 6,500 were "foreigners." But by the end of 1849 no fewer than 100,000 Argonauts, as the gold seekers were called, had descended on the territory. They came from everywhere. The average American prospector in the early rush days was not a grizzled old sourdough but a man in his 20s, educated and reasonably well off. The prospectors also included a large number were immigrants, poor and often illiterate in any language.

For many, California's gold offered the only way to a better life. A poignant extract from a miner's letter to his wife in 1852 expresses the sentiments of many lonely prospectors:

"Jane I left you and them boys for no other reason than this to come here to procure a little property by the sweat of my brow so that we could have a place of our own that I mite not be a dog for other people any longer.... there are murders committed about every day on the account of licker and gambling but I have not bought a glass of licker since I left home.... I never knew what it was to leave home til I left a wife and children.... I know you feel lonsom when night appears but let us think that it is for the best so to be and do the best we can for two years or so and I hope Jane that we shall be reworded for so doing and meet in a famely sircal once more. that is my prayer."

Thousands of every culture, class and calling found their way to the gold fields. Mormons heedless of patriarch Brigham Young's command to stay home rallied to the cry of "California! To California! To the Gold of Ophir!" Men came from Hawaii (the Sandwich Islands), from Europe and Latin America. They came from Great Britain and the colonies. Men like the famed African explorer Henry Morton Stanley mingled with jailbirds from the convict colony of Australia. The clannish French who came in large numbers were referred to as "Keskydees," because their inevitable reply to a question in English was a puzzled "Qu'est- ce qui'il dit?. By 1852 there were 25,000 Chinese in the state who were the object of discrimination and often restricted to sifting another man's leavings for minute bits of overlooked gold. There were 120,000 miners digging in the California foothills at the peak of the gold rush. Twenty years later there were only 30,000, half of them the persistent Chinese gleaning worked-over gravels.

More Chinese came to the gold fields than any other foreign group. Not all who set out for the Golden Land made it. Some fell victim to unscrupulous countrymen who lured them from their villages to Chinese ports, bilked them and disappeared. One group of 800 Chinese who sailed for California in the mid-19th century ended up as slaves digging guano on the scorching, Godforsaken Chincha Islands off the coast of Peru.

Not Many Women

Not many women went to the gold fields. Those that did faced a rough life in the raw tumult of the mining camps. The California census of 1850 lists less than eight per cent of the population as female. Some of the women who followed their men were hard workers like Mrs. R., who, according to an admiring miner, was a "magnificent woman!...earns her old man $900 in nine weeks taking in washing." The gold fields attracted a handful of earnest and unsuccessful temperance lecturers like a Miss Farnham who conceived a scheme to end the pervasive lawlessness of the frontier by the importation of "5,000 virtuous New England women."

Women were indeed imported; not all were virtuous. An industrious prostitute could easily make a fortune. One young woman retired after a year's activity which netted her $50,000. On the other hand unsuspecting working girls back East who were hired as "domestics" sometimes found themselves in bordellos and gambling halls. More of them than Miss Farnham would have admitted made the best of it, taking advantage of a market in which an attractive girl could name her price. Men fresh from the rigors of the camps were starved for entertainment. They showered bags of gold dust on actresses and dancers. In the remote camps a woman's well worn satin slipper could bring $50.

In the early days gold dust circulated as currency along with coins of every nation and coins and stamped ingots issued by various private mints. Men carried a leather poke to hold their gold and a miniature balance scale to weigh it. A pinch was worth about 25 cents. An ounce of gold was valued at $16 but often brought less. It was not uncommon for adulterated gold dust or bogus "retort nuggets" alloyed with base metals to be fobbed off on the unwary. A determined prostitute from Hong Kong once took a client to court in San Francisco because he had paid her in watered gold dust. She won her case!

Ladies of "virtue" were very much in demand. One woman, of this scarce breed buried her husband on one day and married the chief mourner the next. Funerals were a common sight. The 49ers were a profane but highly superstitious lot, and they never failed to say a few words over a fallen comrade. One cloudy day a group of men bowed their heads around a freshly dug grave as a preacher intoned a prayer. Suddenly, the sun came out lighting up flecks of gold, spangling the newly fumed earth at their feet. Every mourner, including the preacher, whooped and immediately fell to his knees scrabbling at the pay dirt, the deceased already forgotten.

In the universal rush for riches it was a foregone conclusion that although the prospectors might not hit pay dirt, the merchants, whores, shipowners and wagon train outfitters would prosper. The journey to the gold fields was long and dangerous whether one went by land or by sea. Profiteers operating at both ends of the trip congratulated themselves on heady profits.

Eager neophytes who had elected the overland route, which covered some of the most hostile terrain in North America, gathered at the chaotic staging points along the Missouri River. They bought "guides" full of misinformation. Trails were where there were none; fodder and water was shown where there was nothing but burning desert; hundreds of miles were lopped off the route with a stroke of the pen. Neophytes invested their precious grubstake in "gold extractors" and other worthless equipment peddled by unscrupulous knaves. They set out with flaming hopes, but the brutal trails across the West were littered with broken wagons, abandoned equipment and the carcasses and skeletons of mules, horses and cattle. The way led across dusty plains, hellish deserts, and freezing mountains, some so steep horses and prairie schooners had to be lowered over cliffs with slings. Rude crosses marked the lonely graves of the more than 5,000 souls who never made it to the Promised Land.

The other, equally harrowing route to the gold fields was by sea. Around South America's Cape Horn was a voyage through 1,800 miles of treacherous seas in foul, overcrowded ships, which took six to eight months until the advent of the swift clipper ships reduced the time to a little over three months. The alternative route was quicker, taking as little as thirty-four days, and accordingly more expensive, but it was certainly no more pleasant. This journey involved sailing from an East Coast port south to Chagres or Aspinwall on the Atlantic coast of the Isthmus of Panama. Once there men who had paid an extortionate $380 for a first-class passage from New York to San Francisco, the great port of the gold rush, had to cough up an additional fee to be guided up the fever-infested Chagres River and then overland by mule through teeming jungle mud to Panama City on the Pacific. Men who chose this route because it was the fastest often found themselves waiting up to eight weeks in the pest hole of Panama City for a berth to San Francisco. One New York man who had sailed with $475 arrived in California with his grubstake reduced by the frustrating six-week wait to a mere $6.

The overlanders who survived the punishment of the elements, Indian attack, hunger, thirst and illness arrived totally exhausted, resources drained. Those who had made the ocean voyage were no better off. Huddled fire-prone shanty towns of lean-tos or tents took them in. Swift-springing settlements appeared along the gold fields: towns with names like Bedbug, Timbuctoo, Rough and Ready,

Second Garrote, Ophir and Hoodoo Bar. Many names mirror their founding fathers — Spanish Flat, Dutch Flat, Chinese Camp (where the great Tong War took place), Cherokee and Hornitos, so-called after the oven-shaped tombs of Mexican settlers who had been forced out of nearby Quartzburg by a "Law and Order Committee." They staked claims and sent to work on the Mother Lode a belt of gold-bearing quartz ranging from a few hundred feet to two miles wide and more than 100 miles long.

There were no laws either to hinder or to protect them. Such a mixture of cultures, class and character inevitably resulted in a volatile brew. Murder and robbery were punished swiftly with lynch mob justice. A Mexican dance hall girl who killed an American in self-defense was hanged before an enthusiastic crowd despite the fact that she was pregnant. A Yankee shopkeeper who remembered the event wrote in a letter: "I always think of the Spanish girl standing on a plank of the bridge, tossing her hat to a friend and putting the rope around her neck, folding her hands and facing death with bravery that shamed us men. And girls," he added wistfully, "were so scarce in those days too."

San Francisco, the warehouse of the gold rush, was founded as Yerba Buena by a Spanish explorer in 1776. It was no more than a dusty village on the bay in 1848. By the end of 1849 more than 40,000 men had passed through and there were 25,000 inhabitants living in ramshackle huts and canvas houses scattered on the hills and shore. The bay was crowded with hundreds of rotting deserted ships, many with cargoes never unloaded. In the 10 years following Marshall's discovery California had produced $55 million in gold and most of it passed through San Francisco. In the 1850s San Francisco was the archetypical boomtown, averaging 30 new houses, two murders, and one fire daily. The fledgling metropolis was razed by fire six times in a year and a half. The city invited pleasure seekers to part with their gold in a thousand gambling dens and more than 500 saloons.

Con men profited handsomely from the gold rush without ever hefting a shovel. Mark Twain is credited with the observation that "a gold or silver mine is nothing more than a hole in the ground with a liar on top." Exaggerated tales were as common as fleas in the gold camps and gambling halls of the old West. Most of them were simply prospectors' braggadocio. But for every simple boaster there was a flimflam man who found the starry-eyed greenhorn easy prey.

A man who had depleted an alluvial claim and wanted to unload it and move on would sometimes chew gold dust with his tobacco and spit into the water where he was panning. This introduced "color" to impress any prospective purchaser who might be within range.

The bunco artists had countless ploys to exercise on gold-struck suckers who believed the precious stuff was everywhere. Some were as crude as throwing a few ounces of gold dust at the bottom of a sterile open digging to clinch a sale. There were many other forms of "salting." A shotgun loaded with bits of dust could be fired at a gravel or rock face so that bits would stick as though they had been there for eons. A prospective buyer would be taken to the claim and encouraged to take a sample for assay. Claims that had once been famous bonanza strikes were especially easy to salt and sell to the unsophisticated prospector who failed to consider they might have become exhausted.

Less ingenuous investors insisted on enlisting the advice of geologists, engineers and assayers before laying out large sums. But even these experts could be hornswoggled. To evaluate a vein, mining engineers and their crews made random shallow grooves along the vein in question, spreading canvas to collect the chips and then placing the samples identified by groove location in canvas bags fastened with a lead seal. Clever swindlers would paint the surface of the area to be sampled with a solution of gold chloride or inject a little gold dust into crevices.

Or a lead-sealed sack of virgin ore could be tampered with. Ingenious salters made counterfeit seals from wax impressions enabling them to open the sacks and add a little gold dust before resealing them. Sometimes a bit of "color" might be sprinkled onto the rough surface of the canvas bags before they were used. Another method of salting was as straightforward as gaining solitary access to the sealed bags, opening a seam, and after "salt" had been put in, sewing the seams up. There were cases of gold chloride solution squirted into sealed bags by hypodermic needles and gold dust blown in through goose quills. Occasionally a wary potential sucker unmasked a swindler by placing bags of ore already tested and known to be sterile in a place where the salter had access to them. The jig was up if they miraculously contained gold when reexamined.

In San Francisco and the other supply centers, inflation devoured a miner's gold. Building lots in San Francisco went almost overnight from $12 to $10,000; bricks cost as much as $1, shovels $10, and eggs $1 apiece. In the more remote camps along the Sierras a sack of flour could cost $100, a barrel of the same $800, an egg $3, and a potato and slice of bread $1 each. Miners purchasing the indispensable picks, shovels and pans in remote areas paid as much as $100 for each. Some even paid as much as 10 ounces of gold for each of these tools.

A man with quick wits could make a killing. A ship captain brought a cargo of stray cats to rat-plagued San Francisco, where they sold quickly for $10 a head. A young wheelbarrow maker named Studebaker and a butcher named Armour launched their empires in the gold rush towns. Almost every conceivable item sold well even at 20 times its purchase price. One merchant, however, who found himself with an oversupply of tin pans priced at 10 cents each, found a way to move them. At night he sprinkled an ounce of gold dust in the streets around his store. Next day, with a great show he swept up a panful of dirt and professed great amazement when gold glittered among the dust. Within minutes he had sold all his 10-cent wash pans for two dollars each.

In the early years many prospectors averaged $30 to $50 a day, but it meant backbreaking work with one's legs in icy water, toiling under blistering sun or bone-chilling rain. There were men who made great strikes but few were lucky or disciplined enough to keep their finds and retire wealthy. A few prospectors found spectacular nuggets, but most of the gold was dust which showed as "color" in a stream, flakes, grains or small nuggets. The famous nugget found at Sutter's Mill weighed less than a quarter of an ounce. The largest piece of native gold found at Carson Hill in 1854 was a mass weighing 195 pounds. The largest true nugget, weighing a magnificent 54 pounds troy, was found at Magalia in 1859. Until a few years ago a celebration was held annually in its honor. Finding a large nugget sometimes had a peculiar effect. A French prospector found a beauty worth over $5,000 one day. The next day he became insane and was shut up in an asylum although his family in France received the income from the sale of the nugget.

For the first time in civilized history men were free to take valuable treasure from the wilderness — free gold for whoever found it. In the first delirious days, the soaring dreams of a real "strike," came true for many amateurs. A man hunting rabbits near Angel's Camp stuck a stick in the ground by a manzanita bush exposing a piece of gold-bearing quartz. He used a knife to dig $700 worth of gold out of it that day. He returned the next with better tools and gouged out an additional $2,000 more, and $7,000 the third day. Three German prospectors exploring on a high tributary of the Feather River used spoons to dig $36,000 of gold flakes out of cracks in the rock. The news of the strike leaked out and men dropped everything to rush to the area, aptly named "Rich Bar." It was so lucrative claims were limited to ten square feet; one pan of dirt easily yielded $1,500 to 2,000 and the record was said to have been a pan worth $2,900. But such phenomenal strikes were characteristic only of the early days when prospectors were working virgin ground. By 1852 the zenith had been reached; the heyday of the man with a dream and a shovel was over.

In the first few years the yield varied widely from one area to another. More than 40 per cent of California's gold was washed from placer deposits of the western Sierra Nevada. At Sutter's Mill men generally made $25 to $30 a day working a rocker. However, in the gravel bars of the North Fork of the American, the Yuba, Trinity, Feather and Stanislaus rivers it was not uncommon for men to find gold worth $500 to $5,000 in a single day. Gold production increased annually to 2 1/2 million ounces in 1851 and peaked at 3 million in 1853. In 1855 the known alluvial deposits began giving out, although the following year the fabulous placers at Colombia, California, yielded the first real colossal amounts of gold, which was not depleted until the early 1860s.

In 1852 miners, realizing that much gold lay trapped in quartz deep beneath the earth, began sinking shafts in the first extensive underground mining of ancient buried river channels. Stamping mills for processing the mined ore were built. Underground tunnels were carved through solid rock. To get at the subterranean gravels, miners sank "coyote holes," deep shafts to lower levels near bedrock with horizontal tunnels or "drifts" radiating out. It was difficult, dangerous, and expensive. Cave- ins frequently resulted from weak framing and many lives were lost. At Grass Valley the workings of the Empire Star Mine plunged a mile deep to a honeycomb of over 200 miles of galleries. Because it required significant capital and sophisticated organization, such mining changed the complexion of California gold production. George Hearst, father of the famous news baron, and Leland Stanford were among the men who made colossal fortunes from hard-rock mine holdings.

In the first flush of discovery the 49ers had been content to skim the cream, the easily collected gold from the streams and adjacent banks of the western Sierra Nevada. The average gold seeker had no knowledge of mining or metallurgy. All he needed was an eye for "color," a strong back, a pan, a shovel and perhaps a cradle. Pan washing requires no more than a shallow iron dish, patience and a deft hand. Gold-bearing sand mixed with water is scooped up. With a practiced move of the wrist the sand or mud swirls to the edge of the pan and is washed away, while the heavy yellow grains collect in the center.

A slightly more specialized piece of equipment is the cradle or rocker used by the 49ers and elsewhere later. It is a wooden trough about six feet long, fitted on wooden rockers in a sloping position. The bottom is fitted with transverse wooden slats or riffles. The auriferous sand is shoveled in through a screen at one end, and as the cradle is rocked a stream of water is flushed through to carry the sand away leaving the gold particles trapped in the riffles.

With time the more ingenious miners devised new methods or revived old techniques to increase their production. They used such ancient and rudimentary techniques as winnowing the ore in baskets as was done in ancient Egypt or trapping gold particles in rough woolen blankets reminiscent of the Black Sea area fleeces. Mercury was used in the amalgamation process. They built sluice boxes including one called the "long tom". It was a long wooden trough shaped like an inverted funnel through which a continuous stream of water ran, very much like those used by the Romans. While one or two men continually shoveled dirt another removed stones and debris which impeded the flow. Heavy gold particles were trapped on a slatted bottom as the waterborne detritus washed away. With such a system a much greater volume of earth could be processed.

The long tom and its more elaborate variations required running water. When an immediate source wasn't handy a system of flumes and ditches was dug to carry water, sometimes from several miles away, to the gold-bearing earth. After 1850, millions of tons of gravel were washed in almost 5,000 miles of channels which veined the landscape. Joint-stock companies were formed to supply the co-operative labor, equipment and capital such an effort required. The individual miner who had neither the inclination nor the means to join in such undertakings often set out by himself. Restless and determined, he prospected through the tangled mountain wilderness, tracing the source of every creek and stream in a search for the Mother Lode, the chief vein from which downstream bits had been eroded.

In 1850 miners, impatient with having to dig into gold-laden hillsides and gorges with pick and shovel, developed a new technique called ground sluicing to handle low-grade gravels near the surface. They dug a narrow gully down the hillside they intended to wash. A flume was directed to the top of the channel and water channeled to rush down it as the men shoveled dirt into the artificial stream. A good deal of gold that passed through it was lost, but fair quantities caught on the rocks and other natural obstructions. Every few weeks the debris that had settled in the natural riffles would be washed through a long tom to separate the gold.

By 1852 this had led to a revival of hydraulic mining such as the Romans used in Spain. Water was piped under pressure up to 30,000 gallons a minute through hoses directed on gold-bearing slopes...hills...even mountains. It was a cheap, efficient, and profitable way to get at the gold but horribly destructive. The California landscape today is dotted with massive jumbles of barren rock and sheared mountains, time-softened to an eerie beauty. They are scars of the raw wounds which defaced the mining districts where hydraulic mining washed away forests, hills and whole mountains. Farm lands were flooded and rivers clogged. Even San Francisco Bay was discolored by hydraulic detritus. This was the chief method of gold production from 1864 until 1884, when the practice was outlawed. Hydraulic mining was controlled by highly capitalized gravel mining companies. The free-lance miner was able to find some gold by hand-processing the tailings from hydraulic operations (they were sometimes quite rich), but it was investors in New York, San Francisco, and London who made the real profits.

15 — *The Prospectors Move On...*

More Booms

old seekers who had made the trek to the rolling foothills of the Sierra Nevada too late to find loose surface gold...who had combed the overgrown ravines, panned hundreds of streams and dug through bedrock without finding a bonanza...were understandably bitter. One recalled that "sudden disappointment on reaching the mines did not only sink the heart but sometimes the minds of the gold seekers."

One man wrote that he did find gold but had to dig through bedrock to get it. He could dig only a dozen pans a day from which he gleaned an ounce of gold. "Out of this," he wrote, "it required about $10 per day to supply my food which was usually beef or pickled pork, hard bread and coffee. By extra economy, I sometimes managed to subsist on $8 per day. Since an ounce of gold was officially priced at $16, he like most prospectors was eking out a marginal living.

Dreams of gold didn't wither easily in spite of disillusion. Many men, having been bitten by the gold bug, moved on to seek Eldorado elsewhere. Some prospected throughout the coastal and Rocky Mountain regions with modest success.

In 1858 a cry of "Gold!, gold!" went up on British Columbia's Fraser River, and during the short-lived boom six out of every hundred persons in California shouldered pick and shovel and trekked north to Canada to join the trappers, traders and Indians who swarmed into the territory, swelling the total to 25,000. The sour-doughs in British Columbia found that the British crown had organized admirably to avoid the lawlessness and confusion of the California camps. The digging was government-controlled and licenses were issued for payment of a monthly fee. Miners were supervised and permission to prospect was revoked if there was rowdiness. The Canadian officials, environmentalists before their time, withdrew permission from miners who polluted streams or needlessly destroyed trees.

The amount of British Columbian gold turned out to be grossly exaggerated. More than 30 other unjustified "rushes" occurred during the 1850s. At the merest whisper of gold, fevered prospectors zeroed in on remote parts of California, Washington, Oregon and even South America.

In 1858 veterans from California's depleted gold fields found rich, shallow deposits of placer gold in Arizona. The landscape of the Southwest was already pocked with sites sporadically worked by Spaniards and Indians in the two and one-half centuries before New Mexico and Arizona became territories. The new mining excitement was ephemeral. In the first rosy flush merchants, traders, gamblers and saloon keepers, who always found more gold from working the miners than from mines, were drawn to the area as by a magnet. Gila City sprang up overnight to accommodate the boom. Then, almost as suddenly, the auriferous earth of the canyons and gulches gave out. Within a week the city was all but deserted.

The slightest hint of a strike drew the peripatetic treasure hunters like flies. They swarmed to Utah, New Mexico, Colorado, Idaho, Nevada and South Dakota. Almost everywhere they encountered disappointment and often recalcitrant Indians. Gold was deposited in some of the earth's least agreeable regions, and the mining frontier of the Southwest was particularly hostile. Vast distances across arid, rough land made transportation and communication extremely difficult. Topography and geology made mining hard labor indeed, but men who had the fever met all challenges.

In 1859 in a dry, sagebrush-covered gully of Washoe County near Carson City, Nevada, two poor Irish prospectors saw a pile of peculiarly colored black dirt in front of a ground squirrel's burrow. Mixed in the dirt was the "damned blue stuff" prospectors had been throwing away for years. The Irishmen had a sample tested. The black earth proved to contain a large amount of electrum; the blue earth was rich in silver. The men, who had been averaging a dollar a day, staked a claim. They called it Ophir and it lived up to its name. It was part of the fabulous Comstock Lode. The famous lode, four miles long, is best known today for its silver, but it was so packed with gold that in 20 years following its discovery it was the country's chief gold producer. The Comstock made a score of men

142

magnates for life and was responsible for the "flash-in-the-pan" wealth of many others.

George Hearst, the California mine speculator and developer, got wind of the Ophir find where gold, albeit pale and worth less per ounce because of silver content, was being taken out by the pound. He sold his California gold mine, borrowed some money and bought an interest along with several other opportunist entrepreneurs. As for the discoverers, McLaughlin sold his share for $3,500 and ultimately had a pauper's funeral. O'Riley held on for a while and then sold for $40,000, far less than the claim was worth. He lost his money in the stock market and died in an asylum. The shrewd Hearst, antithesis of the starry-eyed prospectors, parlayed his interest in the Ophir into a fortune, which he then used to involve himself in almost all the major lode-mining operations of the West.

In the Comstock mines Hearst's engineers were faced with age-old problems endemic to deep mining. The techniques they developed to cope with ventilation and ground support problems are still employed today. First they tried methods Agricola had described such as carving out rock supports to shore up the slopes, or rooms; but these collapsed because the ore body was too soft. Subsequently timber-slope mining was introduced in which square-set timbers were used to brace the walls and roofs of a stope as the ore was removed. The wooden cages could be piled on top of each other, the roof of one serving as the floor of another as the miners worked their way upward. Waste rock was emptied below to fill the area already worked.

The sourdough prospector...the independent Argonaut...could find little scope for his dream in mines where he was only a wage laborer. His pans and cradle were of no use 3,000 feet deep where muckers toiled on 15-minute shifts, always fearful of being scalded by steaming subterranean water at 170 degrees. By 1880 Nevada's Washoe Boom was over. The capitalists who had made their fortunes withdrew to San Francisco mansions. The miners, both greenhorns and veterans of earlier strikes, moved on, leaving Virginia City as a monument to the 19th-century boom-and-bust phenomenon. "The miners were like quicksilver," wrote H. H. Bancroft in his history of the Northwest. "A mass of them dropped in any locality and broke off into individual globules, and ran off after any atom of gold in their vicinity. They stayed nowhere longer than the gold attracted them."

143

'Pikes Peak or Bust'

In 1859 the infamous "Pike's Peak" rush got under way as 100,000 sanguine prospectors streamed across the vast stretches of the Great Plains toward the Rocky Mountain gold fields. Fewer than half stuck it out and reached the Rockies. The rush was set off by the rumor of a solid mass of gold weighing tens of millions of tons deep in the bowels of the mountains, pieces of which had been exposed by erosion, broken off and washed into the streams of the eastern slopes. Such ridiculous claims were solemnly printed in Midwestern and Eastern newspapers and fervently believed by the masses who set out in ox-drawn prairie schooners with the cry "Pike's Peak or bust!" ringing in the air.

Pike's Peak was a superb example of promotion. There was some gold at the end of the trail, but not much — certainly not enough to support the overblown expectations of the thousands who had read highly colored newspaper accounts and deliberately misleading guidebooks. The rumor was fed by the string of towns along the Missouri River up from St. Louis. The financial panic of 1857, one of the nation's most severe depressions, had left them in dire straits. A gold rush would boost the economy nicely, and so the actual discovery of a gold pocket on Cherry Creek near the present site of Denver, by a veteran of both the Georgia and California gold fields, was inflated out of all proportion.

The outfitters in the towns of the Kansas and Missouri frontiers rubbed their hands as the newspapers manufactured banner head-lines proclaiming "The New Golconda." Callous promoters published dozens of guides like those which earlier misled the 49ers. They promised gold in every shovelful of dirt and twisted the truth, inventing bridges where there were none, promising roads

Gold offers magnificent opportunities to talented craftsmen such as those who created this Spanish rosary, top, and gold bracelet, below.

as smooth as silk, and ample water and grass for the oxen all along the route. They shortened the journey by hundreds of miles and took poetic license to transform brutal arid desert into lush grassland. In actuality the trail was long and arduous, stretches of it were largely unexplored, water holes were dry and at times great walls of fire swept over the parched prairies. Hunger, thirst and exhaustion were the companions of many who reached Pike's Peak only to find that they had been flimflammed. Many others fell by the wayside as provisions ran out, and there were recorded instances of cannibalism.

Within a year the bubble burst and the bitterly disappointed miners, who had been badly fleeced, threatened "burning down all the towns on the Missouri." Most of the disillusioned returned home but some stayed on, with Denver as their supply center, to find gold in such places as Gregory Gulch the richest of the mini-strikes, Idaho Springs and Gold Hill. Enough gold was produced in this Rocky Mountain area that by 1866 Denver merited a United States mint.

In 1892 the richest gold mining area in the Rockies was discovered at Cripple Creek just southwest of Pike's Peak. A treasure-hunting carpenter who had been prospecting on weekends for 20 years tested a rock called sylvanite. It was so common it was used in construction but his assay discovered it to be a telluride of gold and silver. He followed the trail of sylvanite to its source at the mouth of an extinct volcano, staked a claim, sold it for $20 million and retired. Within months a town with a 120-room hotel appeared at Cripple Creek attracting crowds of treasure hunters and scoundrels. "Crime in our fair city," wrote the marshal of Colorado Springs, "is at an all time low. All the criminals have moved to Cripple Creek." One mine after another was sunk, and by the end of 1896 nearly $22 million worth of gold had been produced. In 1900 population of the district burgeoned to 50,000, and production accounted for almost a quarter of

Location of some of today's gold is shown in this photo of solid gold bars taken in the Federal Mint in New York City.

the country's gold supply. It continued to pour out gold until the 1960s, employing upward of six thousand men during peak periods.

Just as during the conquistadors' gold rushes, it was always the native population that suffered most. "You are already taking our country from us fast enough," an Arizona Indian told J. Ross Browne, "we will soon have no place of safety left. If we show you where these yellow stones are, you will come there in thousands and drive us away and kill us." And that is what happened.

The gold seekers were the first into the ancestral lands of the native Americans. As the trail they blazed was followed by stockmen and farmers, Indians gradually lost their lands and holy places. Despite encroaching on tribal lands the early treasure hunters were tolerated by the Indians. A lone prospector roaming the hills made little difference in the wide-open spaces of the West. But towns were born and grew into permanent settlements attracting increasing numbers of men, women, traders, farmers and ranchers. Then the railroad cut across the country. The native found his land being devoured by the spreading white man's world and his hunting grounds closed to him. The Indians protested in the only way that seemed to draw attention. They attacked the Argonauts' wagon trains and mining camps.

South Dakota's Black Hills

Response by government was sharp and shameful as one late example amply shows. The massive forest-carpeted Black Hills of South Dakota were the Holy Wilderness of the Sioux, their sacred hunting grounds. There were rumors among the white men that the home of the thunder god was filled with gold. The Sioux had found gold but tried to keep it secret, aware of the white man;s lust for it.

In 1874 General George A. Custer with the 7th Cavalry and a force of geologists, prospectors, assayers and reporters entered the sacred Black Hills for the purpose of "exploration." This act violated the Sioux-United States treaty forbidding prospecting or settlement by non-native Americans in the territory of the Dakota Nation. The men of Custer's ridiculously large expedition found ample evidence of gold in the rushing streams and cataracts of the primeval hills.

When Custer returned from what had effectively been a pleasant three-month vacation spent panning for gold, collecting butterflies, hunting elk and pressing flowers at government expense, he reported that gold had been found but that "the hasty examination [although three months would seem to be time for more than a cursory investigation] we were forced to make did not enable us to determine in any satisfactory degree the richness or extent of the gold deposits in that region." Despite his moderate assessment, the nation received word of the gold in Dakota with eagerness.

For a while the government forbade entrance into the Dakota Nation. But even as the army attempted to turn them back companies of miners, organized in Eastern and Midwestern cities, managed to get past. The national economy had been crippled by the panic of 1873, which came after two years of wild railroad speculation, and the government was under great pressure to somehow make gold in the Black Hills available to "Americans." In 1875 a group of Sioux chiefs were brought to Washington in an effort to have them sell the Black Hills. When they refused, the government organized a tribal council of 20,000 Sioux requesting them to relinquish their sacred lands. Of course the council refused.

They were then ordered off the lands and back to reservations on pain of being treated as hostile enemies. Only a few went home. The government launched an offensive against the rest, primarily the Sioux and Cheyenne who were led by the great strategist Chief Sitting Bull. The Indians defeated the United States forces at the Battle of the Rosebud and then wiped out General Custer and his famous 7th Cavalry at the Battle of Little Big Horn. But they were hollow victories, for as the harsh winter approached and the Indians looked forward to returning to the reservations, the government announced that it would refuse to feed them unless they signed over the Black Hills. They did so on September 6, 1876.

The Manuel brothers, Moses and Fred, veterans of gold fields in six other states and Alaska, were among the thousands who swarmed into the narrow defiles and canyons of the new bonanza area. The Manuels located a lucrative placer deposit which they worked through the snowy winter. After the spring thaw they traced it to its source and found a quartz outcrop laden with gold. They took out $5,000 in gold and then sold the claim in 1878 for $70,000 to the Homestake Mining Company, which George Hearst had formed with two partners in California the year before.

The Homestake Company bought up claim after claim as the easy placer gold dwindled, parlaying the company into the largest gold mining firm in the Western Hemisphere. By the mid 1880s the starry-eyed prospectors and their cradles had departed with the last of the alluvial wealth and were replaced by cadres of hard-rock miners working for the Homestake Company. By 1924 the Homestake accounted for about 90 per cent of the total of more than $230 million in gold extracted from the Black Hills since 1876. The 200 miles of workings on 34 levels still yields an average of 400,000 ounces annually.

The United States ranks fourth in world production of newly mined gold today. The Homestake is the nation's largest producer, processing a ton of ore, brought up from mile-deep galleries, to produce a lump of refined gold about the size of a gumdrop weighing three tenths of an ounce. The ghostly shades of the lone prospectors who made the American gold rush one of mankind's most exciting epics pale in the roar of the giant machinery at the great industrialized mines. But in the streams, rivers, canyons and hills of the Southwest ghosts of 19th-century gold seekers have recently been joined by legions of men, women and children who have turned weekend prospecting into a pleasurable and sometimes profitable pastime.

Australia

One of the 49ers who prospected in California was an Australian, Edward Hammond Hargraves. He wrote home that the land on which the gold was found looked much like the countryside of New South Wales. In 1850 he returned to Australia with no fortune but a great deal of experience. He confidently predicted that he would find gold within a week of setting out on a prospecting expedition. True to his word he found gold on Australia's Macquarie River near Bathurst in New South Wales in February of 1851.

The discovery of gold in Australia helped make London the center of the world gold market. Much of the gold from California and Australia accrued to industrial England, the universal creditor as the world's leading 19th century power. The gold reserve of the Bank of England soared from 12.8 million pounds in 1848 to 20 million in 1852. Eighty per cent of the Australian production was channeled through the London market. Until the first decade of this century when production declined sharply, an average annual yield of almost 230,000 pounds of gold was taken from Australia.

South Africa

The site of the next great treasure find was Africa, the vast continent which had produced such a wealth of gold for the Egyptians, Phoenicians, Romans, Byzantines, Arabs and of course the tribal Africans. This 19th-century treasure, however, was not gold but diamonds, discovered in large numbers at Kimberley along South Africa's Vaal River in 1867. The diamond fields drew thousands of starry-eyed men who easily transferred their passion from glowing gold to sparkling diamonds. Six tons were mined in the two decades following the opening of the Kimberley fields.

While some men searched for diamonds others spread out exploring for gold, finding some alluvial deposits in the Transvaal, which were worked in a random manner. But these discoveries attracted little attention in the excitement of the fabulous diamond pits. Then in 1886 came the crucial discovery, one that was to lead to the world's greatest source of gold, a stratum of gold ore deposited eons ago when the sea lapped at the long ridge of the White Waters, Witwatersrand, in central South Africa.

Gold seekers poured into the Transvaal following the perennial dream of a bonanza which had swept some of them back and forth across oceans and continents since 1849. The early acts of the South African drama followed a now familiar course: discovery, influx of prospectors, overnight mushrooming shanty town (this time supervised from the outset by government army officers) and, finally, the withdrawal of the protagonists with scant reward.

But from that point on the script was radically different. This gold rush was to be unlike any other. There was no place for the lone miner. The nature of the reef, which curves in a wide arc 300 miles east and southeast of Johannesburg, required gold to get gold. There were few outcrops of easily accessible gold ore; most of it lies in bands varying in thickness from one tenth of an inch to 100 feet, averaging no more than a foot, all deeply buried beneath thousands of feet of rock.

Massive capital was needed to process the rich ore brought up from veins which inclined steeply into the bowels of the earth. At first the crushed ore, treated with mercury, yielded only about 75 per cent of the gold it contained. It was discovered that ore brought up from more than 150 feet below the surface was refractory. The gold was enclosed in "sulpherets," metallic sulphides which prevented mercury from releasing all of the gold from its matrix. No matter how finely it was milled a great deal of gold was lost, making the mining effort unprofitable, since a ton of crushed ore yielded only one-third ounce of gold and sometimes less. The huge volume of tailings was sold to make gold-filled cement for the growing city of Johannesburg.

South Africa's gold industry was saved by scientists in Glasgow, Scotland, who developed the cyanide process of extracting gold from crushed conglomerate. In 1890 the first cyanide treatment plants were set up in the Rand. Deeper mine shafts were sunk, and by 1892 South Africa was supplying over one million ounces of newly mined gold a year; more than 15 per cent of the world's production. In 1898, the year before the growing conflict between the farming Boers and the mine-oriented Uitlanders, outsiders, erupted in the bitter Boer War, South Africa out-produced the United States for the first time. The Rand mines that year yielded a total of 4 million ounces, more than 25 per cent of the world's gold production.

The Boer War interrupted mining for three years. The British victory in 1902 led to equalization of the British-Afrikaaner population and the formation of the Union of South Africa in 1910. Mining was resumed on a massive scale largely controlled, as it is today, by the seven major mining houses which make up the Chamber of Mines of South Africa. Since 1884 the mines of the Witwatersrand basin have contributed an average of more than 49 per cent of the free world's annual newly mined gold. In 1990 South Africa produced 605.4 metric tons of gold; USA produced 295.0; the Soviet Union 260.0; Australia 241.3; and Canada 165.0. Overall, South Africa has accounted for about 40 per cent of all the gold produced in history.

16 — *The Final Stampede...*

Klondike

Alaska gave the gold seekers one final rainbow to follow. There was treasure at the end of it but also more suffering and sorrow than the eternally optimistic Argonauts had ever encountered before. The last of the 19th-century gold rushes, the epic Klondike Stampede, was set off by a thumb-sized nugget found on a tributary of the Yukon River, as close to the North Pole as northern Siberia.

The stream bore an Indian name meaning "abundant fish" which became "Klondike" in the mouths of the prospectors. Few men ventured into such a remote wilderness area of moose pasture, mountains and rivers that were frozen from September to May. Even fewer lived there. But an American, George Washington Carmack, born in a covered wagon crossing the Great Plains and the son of a failed 49er, had come north in 1855 to prospect. He stayed on to marry a Tagish Indian chief's daughter. "Siwash George," the "squaw man," as he was derisively called by the few sourdoughs who prospected the area, was more interested in salmon fishing and moose hunting than gold. He was considered a strange man, a lover of Indian life who read *Scientific American*, had an organ in his cabin and composed sentimental couplets. One night George had a visionary dream in which a huge king salmon shooting up the rapids stood on its tail before him. The fish's scales were golden and its eyes were $20 gold pieces.

The next morning he went fishing for salmon on the Trondiuck River. On the way Siwash George and his two brothers-in-law, Tagish Charley and Skookum Jim, encountered Robert Henderson, a grizzled old sourdough from Nova Scotia who had spent his life seeking gold first in Australia and New Zealand, then trying his luck across the world in the Rocky Mountain states. Finally he had been borne along with the north-flowing tide to Alaska. He told Carmack of a likely area he had just found but not yet explored. It was a creek that drained off the Dome, the highest mountain thereabouts. Carmack was welcome to stake a claim there..."but I don't want any Siwashes staking on that creek," he added. His anti-Indian remark cost him a fortune, for a few days later the squaw man and his brothers-in-law went up to Rabbit Creek and on August 17, 1896, found the nugget that started the Klondike Odyssey. They staked claims.

The accounts are not clear but probably because of Henderson's remark the three men didn't keep a verbal agreement to bring him in for a share if they found anything. On the way home they passed Henderson without mentioning their find but told everybody else they met.

The word flashed up and down the Yukon Valley. The settlement of Fortymile became a ghost town overnight. Camps scattered along the wilderness streams emptied as the old-timers once again responded to the cry of gold. Where moose had pastured only days before, a town sprung up called Dawson City. Men straggled in every day eager to stake a claim. The area along Rabbit Creek, renamed Bonanza, was quickly taken up and latecomers, called *cheechakos* , probed other streams flowing off the Dome. Five lucky men staked claims on one of these rivulets. Spurned by the more experienced miners as the Bonanza "pups," they eventually produced a million dollars in gold. One of the men drawn to the gold strike was a barber from Circle City whose claim yielded $50,000 a year for five years. During those first months, before word had spread across the sparsely settled Northern wilderness, some men reaped rich rewards; the average take at first, by one estimate, was about $850 a day. But they paid a high price in suffering and deprivation.

The following summer, after the thaw, two ships reached Seattle and San Francisco bringing three tons of ore and a number of frostbitten sourdoughs who had made enough to retire rich. In Seattle the sight of men minus noses, fingers and ears didn't dampen the enthusiasm of a crowd of 5,000 that watched the men escorted by Wells Fargo guards debark with their gold in moosehide pokes. Three hours after the *Portland* had docked at Seattle the waterfront was teeming with men shoving and jostling in an effort to book passage north. Merchants, lawyers, pimps, teamsters, bankers, ministers and even a number of enterprising "box-house" girls...everyone wanted to go. "Gold! Gold!" trumpeted headlines across the country. Within 10 days 1,500 people sailed from Seattle, paying $1,000 each for tickets. The stampede was on.

Two routes led to the Klondike. The easier, more expensive way into Alaska's interior and the forbidding Yukon wilderness was the all-water voyage up the Pacific to the Bering Sea and thence 2,300 miles up the Yukon by steam-wheeler to Dawson. However, the Bering Sea at St. Michael is frozen from late September to late June. The trip, depending on whether a ship got caught in the ice, could last from forty days to eight months and thus was either the quickest or the slowest way to the gold fields.

Many more men bought passage on dangerously overcrowded vessels as far as Juneau, Skagway or Dyea and then trekked 600 miles over Chilkoot or White Pass. Sled dogs were snapped up at $250 and more apiece and soon gone. Horses, glue factory material brought from the States by profiteers, fell on the trail and rotted. Of the 3,000 horses that set out from Skagway to cross White Pass in 1897 no more than a dozen survived. Major J. M. Walsh, sent as Commissioner of the Yukon, wrote, "such a scene of havoc and destruction...can scarcely be imagined. Thousands of pack animals lie dead along the way, sometimes in bunches under cliffs, with pack saddles and packs where they have fallen from the rock above.... The inhumanity which this trail has been witness to, the heartbreak and suffering which so many have undergone, cannot be imagined."

Jack London also chronicled the ghastly scene:

"The horses died like mosquitoes in the first frost and from Skagway to Bennett they rotted in heaps. They died at the rocks, they were poisoned at the summit and they starved at the lakes; they fell off the trail, what there was of it and they went through it; in the river they drowned under their loads or were smashed to pieces against the boulders; they snapped their legs in the crevices and broke their backs falling backwards with their packs; in the sloughs they sank from fright or smothered in the slime; they were disemboweled in the bogs where the corduroy logs turned end up in the mud; men shot them and worked them to death and when they were gone went back to the beach and bought more. Some did not bother to shoot them, stripping the saddles off and the shoes and leaving them where they fell. Their hearts turned to stone...those which did not break, and they became beasts...the men on the Dead Horse Trail."

Men generally fared a little better, although of the 100,000 who set out for Dawson only 30,000 to 40,000 arrived and, according to Pierre Berton in his excellent book *Klondike* (London: W. H. Allen, 1960), only half of those who reached the "Paris of Alaska" bothered to prospect. Perhaps four or five thousand found gold, but only a few hundred realized their golden dreams, becoming really rich.

The brutal mountain passes were hellish at any time of the year, and yet in the winter of '97 more than 30,000 Argonauts hopelessly in thrall to the age-old desire for shining gold made it over the sheer walls of ice, forcing themselves ahead step by slogging step in a line that had no end. It was an appalling spectacle, recalling the forced marches of slave armies to the Egyptian desert mines so many thousand years before. It seemed incredible that these 19th-century men would willingly undertake such a difficult and perilous journey — no matter how much their lust for gold.

Most were eventually forced to pack anything they would need on their backs. Shifting snow on the high passes drove them to their hands and knees as a relentless wind flung splinters of shattered ice to lacerate any exposed flesh. Temperatures fell to forty below. Sweat froze, binding clothing to the skin with one step and ripping the frozen skin away with the next. Blizzards blinded and glacier ice toppled. Men without a year's supply of food were turned back by the Canadian Northwest Mounted Police in scarlet jackets and white helmets who were posted atop the passes in an effort to prevent disaster. Once over the summits the men pushed on to Lake Bennett. They built crude rafts from green lumber as members of the police circulated among them warning of the dangerous rapids, reefs and narrows that lay ahead for 500 miles, advising the weary men to build their craft long and strong. When the Yukon thawed, the police counted 7,124 boats ready to depart. Most made it; crosses on the banks below White Horse Rapids testify to those who didn't.

There was an even crueler route to Dawson, the sinister Ashcroft trail which wound through the dripping wilds of British Columbia. At least 1,500 gold seekers and 3,000 pack animals attempted the journey. They found a desolate land bare of fodder, a terrain of bottomless bogs where they suffered unremitting torture from mosquitoes and flies. In rainy fir forests, dark even in daylight, putrid carcasses of animals marked the trail. Men drowned, were murdered, died of hunger and fever or blew their own brains out in desperation.

The embryo town of Dawson, 4,000 miles from civilization, was crushed beneath the 30,000 sourdoughs who crowded in by spring of '98. If a man lost sight of his partner, he might not find him again for days. Exhausted but sanguine gold seekers arrived daily. Dawson had just come through nine frozen months of starvation, scurvy and typhoid. During that time the police would admit no one to the town's jail unless he had his own beans and provisions. Food was shared and rationed. Women sold themselves to the highest bidder for enough to eat and men lay muffled in blankets during most of the long gray days.

In the spring Dawson suddenly became a bazaar by day and a frenzied carnival at night. After the thaw fresh supplies arrived along with throngs of cheechakos and sourdoughs. Among the tents and plank shacks were two banks, two newspapers, five churches and dozens of saloons, dancing halls and gaming houses. The currency of every country, even Confederate notes, competed with gold dust to buy champagne, fresh grapes, oysters, ostrich feathers and opera glasses. Within two years there was a telegraph, steam heat, electricity and the railroad. Thanks to the calm authority of the Mounties reasonable order was maintained throughout. The police commander ordered the San Francisco dancing girls to change their bloomers for more modest skirts and was obeyed. The closest thing to a murder Dawson City saw in 1898 was an altercation between Coatless Curly Munro and his wife. In the heat of the fight they each went for revolvers they kept under the bed pillows but fled the room by separate doors before firing.

In 1899 the sourdoughs wrested some 5,000 pounds of bright gold from the frozen ground. By 1900 a peak of 45,000 pounds was reached and the stampede was over, although commercial mining continued to be profitable until 1966.

Robert Service, poet laureate of the Klondike, wrote eloquently of the trail of '98, of the "big shiny nuggets like plums" that danced before the eyes and of the men on the tortured trek to Dawson City. Little wonder that so many of those who survived didn't even bother to look for the shiny stuff once they reached there. Despite the fact that Siwash George and other sourdoughs first on the scene had washed abundant gold from the streams and rivers, the Klondike was not nugget country.

It quickly became apparent there was only one way to get the most out of a claim — by digging, which was slow, agonizing work. The gold lay in a gravel matrix, in benches and shelves, buried five to 30 feet beneath heavy, boggy tundra. During the long, dark winters, when the temperature hovered around minus 60 degrees Fahrenheit, a shroud of smoke hung over the diggings as men used fire to soften the ground, doggedly hacking through the permafrost. To get wood for the fires they logged the streams or struggled to bring felled trees from the forested mountainsides. They lit the fires at night and dug by day. A foot a day was the average rate of descent to bedrock. The cold was bitter and insistent. Men spread a thick layer of bacon grease mixed with ashes on their faces to protect them from frostbite. At 40 below the mercury froze in thermometers and at 80 below ax-blades shattered. Even their Hudson Bay rum froze as sourdoughs huddled in tiny cabins. In a desperate bid to warm their bones they fed the fire with chairs and sluice boxes...anything to keep it alive. Occasionally a man fell prey to "cabin fever" and wandered off to a crazed death in the frozen expanses of snow and ice.

When streams thawed in the spring, there was water for sluicing the great piles of dirty gravel that had mounted up over the winter. Pay dirt was washed under the eerie midnight sun.

Some men found that frostbite, gangrene, loneliness and hardship had paid off handsomely. A relative handful became overnight millionaires. Yet within a few years most of these had been parted from their wealth. Drink, gambling and women accounted for the loss of a lot of the hard-won gold. Charley Anderson's experiences are typical. Anderson, the "Lucky Swede" who had bought a million-dollar claim for $800, ultimately lost his dance-hall wife and his fortune which had been heavily invested in earthquake-prone San Francisco real estate. He died in 1939 pushing a wheelbarrow in a British Columbia sawmill. But until the day he died he wholeheartedly believed that someday soon he would strike it rich again.

Prospectors were notoriously poor businessmen. The Croesuses of the Klondike went on to lose their gold in dubious mining investments elsewhere or went bankrupt investing in real estate, railroad and banks. It was not surprising in the light of the extraordinary circumstances surrounding the last 19th-century gold rush that there was a fair share of madness and suicide among the winners as well as the losers of the Klondike.

Contrary to precedent, however, the man who started it all ended his days well off. Carmack made and kept a fortune and even wrote an account of the whole saga. His Indian wife fared less well. He abandoned her in 1900 and she died on a reservation in 1917 wearing a cheap cotton dress. Her only Klondike souvenir was a necklace of nuggets taken out of Bonanza Creek. Carmack married again. His second wife had run a "cigar store," a bawdy house in Dawson where business was so good she had been able to glean about $30 in gold dust from the sawdust on the bar floor every morning. Robert Henderson, the inveterate and embittered prospector, who felt that he had been cheated by Carmack, was recognized as the co-founder of the Klondike and awarded a Canadian government pension of $200 dollars a month. He never ceased looking for gold and died of cancer in 1933 still talking of the bonanza he was going to find over the next hill.

Veterans of the Klondike included men of character and enterprise who made their fortunes later. Among them were Augustus Mack, founder the Mack automobile company; Sid Grauman, who built Hollywood's Chinese Theater; Key Pittman, Nevada senator and chairman of the Foreign Relations Committee; and the three Mizner brothers.

No one who took part in the amazing saga of the Klondike came through unchanged. As Pierre Berton noted, that singular era produced men and women of every conceivable background who were joined by a common bond. Many of them were individuals of unusual mettle, wise beyond their years, who had been tried and proven. They were survivors, achievers whose lives and characters were shaped by the Klondike experience.

Within three years of Carmack's discovery the stampede was over. Dawson, like most other mining camps, emptied as word of fortunes to be made on Nome's beaches reached town. In one week in August 8,000 persons left Dawson. Men moved on to other new strikes which followed at Nome, Fairbanks, Keno Hill and Atlin. The day of the 49er, the fossicker, the sourdough, was over.

At Nome, on the Bering Sea just across from Siberia, men panned over a million dollars of fine gold dust (at sixteen dollars an ounce) from the rich, blue beach sands in the first two months following the strike in 1899. When the gold gave out on the shore, thousands of men moved inland to the hills and gullies with their pans, rockers and wheelbarrows. The beaches today are still littered with rusted machinery left by the sourdoughs. One group of seven men took out $750,000 in the first season. But again, as in the Yukon, because most of the gold lay in subterranean gravel, mining syndicates, financed out of San Francisco, London and New York, soon gained control. They worked the gold fields with huge hydraulic machines and dredges which in one day equaled the labor of a thousand sourdoughs. By 1907 the tide turned and Alaska retreated into gold-rush oblivion.

Conclusion

Today's Gold

Gold was the great population builder for the United States in the 19th century, the opener of new country. The Great American West was first settled by men and women looking for gold along with the legions who followed to supply them. Homesteaders, ranchers and farmers came later, but only after prospectors had opened the land. Gold built railroads, cities, states and territories as well as fortunes and character. From the gold rushes came inventions and improvements in mining technology and engineering as well as countless fascinating stories of unusual men and women.

Economically, the future of gold production in the western United States is dim today. Almost all of the gold and silver now being mined is a by-product of base minerals. To make a profit gold mining concerns in Nevada, for example, have to process seven tons of ore just to recover one ounce of gold. And, the gold particles they recover are so minute they can be seen only with an electron microscope.

In the century and one-half since the great California gold rush the value of all the gold and silver mined in the West, excluding Alaska, probably totals little more than $30 billion. That's a lot of money, true. But, that "lot of money" is actually less than three years worth of the annual agricultural production for that same geographic area!

Yet it wasn't so much the actual economic worth of the treasure that made it significant as its profound and far-reaching effects...psychologically and spiritually. The era began as a free-for-all for every man and ended as a corporate take-over. Too many men and women quested after too little gold in the explosive decades. Prospectors, self-hypnotized by dreams of fame and fortune, darted from one rumored strike to another followed by the inevitable swindlers, profiteers, developers and stock manipulators.

Finally, during the Great Depression of the 1930s when this nation had little to offer but hopes and dreams, as many as 15,000 Americans returned to the Sierra Nevada streams and foothills, abandoned for almost 75 years, to search for gold.

Tourists today explore the crumbling ghost towns with their mountainous piles of tailings and rusted hulks of machinery, trying to recapture the feeling of the 19th century when men were made and unmade by inert shining metal. Weekend prospectors with pans, metal detectors and portable sluices comb Western streams and rivers. Scuba divers in the Caribbean and Gulf of Mexico seek the remains of 16th- and 17-century Spanish ships whose plundered wealth never left the Western Hemisphere. Hobbyists scan the surf and beaches with detectors, searching for the same treasures. Families pan for gold and countless men and women...boys and girls...all eagerly track lost mines and buried treasure.

The dream dies hard. Human nature hasn't changed and probably never will.

By Charles Garrett

Appendix

J enifer Marx has gathered so much knowledge about gold and has compiled it into such an absorbing story that I wonder how it was possible for her to accomplish this task. This may be the most fascinating — but, true — account of man's exploits ever written. It rivals the histories of Alexander the Great, Marco Polo, the Crusades and the like...even the story of man's taming of the American West. Indeed, it *tells* much of the story of the American West.

It's an honor for me to write this appendix. I appreciate the opportunity to praise this fine book, but I also want to explain how gold has been important to both Ram Publishing and Garrett Metal Detector Companies since their earliest days. Garrett detectors have made some of the world's most important gold nugget discoveries, and Ram books and Garrett videos have provided instructions on searching for gold with modern-day equipment. Of course, Garrett manufactures and markets the world's most popular gold pan with its patented Gravity Trap® riffles. In addition, we sell an affordable Gold Panning Kit that includes everything needed to pan for gold successfully.

For more than four decades — from 1933 until 1974 — citizens of the United States were prohibited by law from owning gold bullion, and gold was not accepted as legal tender in monetary exchange. These were among the measures inaugurated by President Franklin D. Roosevelt in the early days of his presidency as he sought to share up the economy of the

163

nation at the depths of the Great Depression. Throughout this period when private ownership of bullion was prohibited, the government set the price of gold at artificially low prices...in 1974 it was still only $35 per ounce.

After the ban on ownership was lifted, the price of gold soared. In the harrowing times of steep inflation and high interest that followed, this price approached $800 per ounce. Such a sharp increase in value, coupled with man's adventurous spirit, caused a small scale gold rush back to the 19th-century camps and ghost towns...gold country of the American West. The art of gold panning flourished once again, with metal detector manufacturers supplying new electronic gold-locating equipment. At the Garrett factory we were in full production, yet deliveries sometimes lagged as much as three months behind sales during the boom.

Although the price of gold has decreased from its highs of the early 1980's, it remains — as this book is written — some ten times greater than it was when the ban on ownership was lifted in 1974. Furthermore, a strong demand for gold continues to exist in the private sector. With its industrial use steadily increasing, prices can be expected to rise generally — especially in relation to international economic conditions.

And, gold is still so beautiful!

Just as the 19th-century prospectors Jenifer describes sought to control their own fate and fortune through the discovery of nature's wealth, so today's seekers after gold pursue similar goals. The spirit of the 49ers and of the Klondike prevails among both professional and recreational prospectors who have discovered the pleasures and profits of searching for gold.

Two current Ram books, *You Can Find Gold with a Metal Detector* and *The New Gold Panning is Easy* describe in layman's language just how simple it is to find gold. Two Garrett videos, *Weekend Prospecting* and *Gold Panning is Easy* depict pictorially how to find gold with metal detector and pan. Both books and videos use simple layman's language that can be easily understood.

The basic instructions are indeed simple:

1. Search where gold is known to exist;

2. Use the right equipment;

3. Be persistent.

Who Are These Modern-Day Gold Seekers?

Doctors, attorneys, businessmen, students, senior citizens...entire families...from all walks of life and all income levels have found that the healthful, relaxed outdoor life of weekend or vacation prospecting can yield dividends — in dollars and cents as well as pure pleasure. Men and women, boys and girls, who search for gold work at regular jobs, but on weekends, holidays and vacations they join their families or friends at one of the many thousands of areas open to the public where gold can be found. They set up camp in the mountains by a stream or in the desert. They use a gold pan, perhaps in conjunction with their metal detector, and they set out to find gold. Or, they may travel to an old ghost town or deserted mining camp and search for nuggets or ore veins or valuable mineral/ore specimens overlooked by the 19th-century prospectors. Those early miners, seeking gold only with eyes and instincts, operated without the benefit of modern electronic equipment.

Modern Equipment

Three developments have greatly increased the ability of the recreational miner to hit paydirt in comparison with sourdough prospectors described in this fine book who flooded California in the 1850s and went to Alaska at the turn of the century:

1. Availability of an easy-to-use, highly efficient gold pan;

2. Production of lightweight, portable dredges that can be effectively operated by one or two people;

3. Development of the metal detector. This single tool makes locating precious metals simpler for all types of prospectors, no matter what their level of experience, no matter where they are searching...on land or in the water. The earliest metal detectors were welcomed just after World War II by prospectors who sought any advantage that would improve their chances of striking it rich. Today's rugged, yet highly sensitive, computerized metal detectors are capable of operation and detection in even the difficult terrain of the most highly mineralized rocks or soil.

On almost all of my excursions into gold country I meet vacationers and weekend holiday fun-seekers who are taking advantage of new developments in equipment to join in the quest for gold. And, I believe that there is nothing more beautiful and breathtaking than gold country itself, whether it be in the high Rocky Mountains or arid deserts. Just being here brings joy. That wealth can be found so easily (relatively speaking!) in such beautiful surroundings is a blessing indeed.

Using a Detector

The ideal metal detector that will enable you to achieve the greatest success for prospecting is a sensitive, deepseeking instrument that has been designed for prospecting and proved in field use. Garrett's Scorpion Gold Stinger and the entire line of Garrett's CX-type detectors are ideal for prospecting. Garrett's amazing new GTI detector with TreasureVision™ offers the additional advantage of informing the treasure hunter about the size of each target that is discovered.

What should such a capable ground-balancing metal detector — one that will find gold — cost? Gold-seeking instruments are available for the equivalent of the value of single ounce of gold. After that, the continuing cost is only a few dollars for replacement batteries. Metal detecting is a one hobby that can truly pay its way.

Complete details on searching for gold with a metal detector are explained in Ram's *You Can Find Gold with a Metal Detector* . All of the other Ram books that include instructions for using a metal detector also describe this most valuable function of the instrument.

It's Easy to Pan for Gold

After complying with the three basic instructions outlined above, follow these general procedures to recover gold successfully by using a pan in a stream of running water:

— Place material suspected of containing gold in your pan;

— Use enough water to keep all material in the pan under water; fully submerge the pan, if enough water is available.

— Run hands through material to thoroughly wet it, top to bottom, and produce a "liquid" state of suspension;

— Rotate the vessel or container under water vigorously in circular motion;

— Remove larger rocks that are washed clean;

— Shake in circular motion, sideways, front to back, up and down (it all achieves the same result);

— Let lighter material "spill" off gradually over the pan's edge.

— Finally, there will be only the heavier material (concentrates) left in the bottom which should be examined carefully for gold.

These instructions are presented with accompanying illustrations in *The New Gold Panning is Easy* .

Successful "dry panning" is also possible; that is, using a pan to recover gold in the desert or any other environment without water. These instructions too are detailed in Ram's book on panning.

Because I know that so many of you who visit gold country want to try your luck in panning for gold, Garrett designed a kit that includes all of the equipment need for panning. It's light and compact, but — most importantly — it's inexpensive! The kit contains the standard 14-inch Gravity Trap gold pan along with a small 10 1/2-inch finishing pan and a classifier (sieve) with uniform, square holes to use in combination with the pans. Also included is a small suction bottle for recovering gold and an instruction booklet that tells all that any beginner needs to know about panning. In fact, this kit can completely serve the needs of both the gold prospector and the rockhound gem hunter.

For additional information on Garrett equipment and where to buy it call 1-800-527-4011. You can use a credit card to call in your order for the gold panning kit or any Ram book or video that will help you in your quest for gold. An order blank for these items can be found on the following page.

Good hunting! I'll see you in the *gold* field...

Form for Ordering...

Ram Books

Please send the following books:

- ☐ Gold of the Americas $12.95
- ☐ Find Gold with a Metal Detector $12.95
- ☐ Gold Panning Is Easy ....$ 9.95
- ☐ Ghost Town Treasures ....$ 9.95
- ☐ Real Gold in Those Golden Years ....$ 9.95
- ☐ Let's Talk Treasure Hunting $14.95
- ☐ Buried Treasures You Can Find $14.95
- ☐ The New Successful Coin Hunting $12.95
- ☐ Modern Metal Detectors $14.95
- ☐ New World Shipwrecks: 1492-1825 ....$16.95
- ☐ Treasure Recovery from Sand & Sea $14.95
- ☐ Sunken Treasure: How to Find It $14.95
- ☐ An Introduction to Metal Detectors ....$ 1.00
  (No shipping/handling charge for this book)
- ☐ Find an Ounce of Gold a Day ....$ 3.00
  (Included free with Garrett Gold Panning Kit)

Gold Panning Kit

- ☐ Complete kit for gold panning $29.95
  (Kit requires NO shipping/handling charge.) Also, when Gold
  Panning Kit is ordered, no shipping/handling charge for any books.

Videos

- ☐ Weekend Prospecting $14.95
- ☐ Gold Panning is Easy $14.95

Ram Publishing Company
P.O. Drawer 38649
Dallas, TX 75238
FAX: 972-494-1881
(Credit Card Orders Only)

Handling charges:
Please add $1 for
each book or video
(maximum of $3)

Total for items $_____

8.25% Tax (Texas residents) $_____

Handling Charge $_____

                        TOTAL $_____

    Enclosed check or money order

    I prefer to order through

☐ MasterCard

☐ Visa

By telephone:
1-800-527-4011   _____

             Credit Card Number

_____

Expiration Date   **Phone Number (8 a.m. to 4 p.m.)**

_____

Signature (Credit Card orders must be signed.)

_____

NAME

_____

ADDRESS (For Shipping)

_____

CITY, STATE, ZIP